Milestones in Immunology

Milestones in Immunology

Based on Collected Papers

Domenico Ribatti
University of Bari Medical School, Bari, Italy
National Cancer Institute “Giovanni Paolo II”, Bari, Italy

Academic Press is an imprint of Elsevier
125 London Wall, London EC2Y 5AS, United Kingdom
525 B Street, Suite 1800, San Diego, CA 92101-4495, United States
50 Hampshire Street, 5th Floor, Cambridge, MA 02139, United States
The Boulevard, Langford Lane, Kidlington, Oxford OX5 1GB, United Kingdom

British Library Cataloguing-in-Publication Data
A catalogue record for this book is available from the British Library

Library of Congress Cataloging-in-Publication Data
A catalog record for this book is available from the Library of Congress

ISBN: 978-0-12-811313-4

Publisher: Sara Tenney
Acquisition Editor: Linda Versteeg-buschman
Editorial Project Manager: Fenton Coulthurst
Production Project Manager: Priya Kumaraguruparan
Cover Designer: MPS

Typeset by MPS Limited, Chennai, India

CONTENTS

ABOUT THE AUTHOR

Domenico Ribatti was born in Andria, Italy, on December 27, 1956. He was awarded his M.D. degree on October 1981. His present position is full Professor of Human Anatomy at the University of Bari Medical School, Italy. He is the author of eight monographs. History of research on tumor angiogenesis, Springer, 2009. The chick embryo chorioallantoic membrane in the study of angiogenesis and metastases, Springer, 2010. Protagonists of medicine, Springer, 2010. Mast cells and tumors. From biology to clinic (in collaboration with E. Crivellato), Springer, 2011. Morphofunctional aspects of tumor microcirculation, Springer, 2012. Angiogenesis and antiangiogenesis in hematological malignancies, Springer, 2014. Development of immunological competence, Springer, 2016. The role of microenvironment in the control of tumor angiogenesis, Springer, 2016.

FOREWORD

A highly discriminatory immune system is fundamental to survive. Ethymologically, the word immune is derived from the Latin legal term "immunis," meaning "not liable for duty," and applied in later Roman times to a group of privileged young men protected by their high social class from the obligation of undertaking the ordinary military service of the state. To this extent, therefore, the expression "protective immunity" would devour of tautology, and in modern usage, the term immune still carries the connotation of protection, as in the many prophylactic immunization measures involving the use of antigenic material that are widely employed in the control of certain epidemic diseases.

The science of immunology grew from the common knowledge that those who survived many of the common infectious diseases, rarely contracted that disease again. Immunology as a distinctive subject developed in the middle of the 20th century as researchers started to understand how the adaptive immune system aids in defense against pathogens. In 1986, David W. Talmage wrote that: "The history of immunology is fascinating in part because the central concepts that power its research have changed so rapidly. In the 30 years before 1948, the word 'lymphocyte' did not appear in the index of 'Journal of Immunology'. Today, half of the papers in that much expanded journal involve research of some aspects of lymphocytes" (Talmage, 1986).

Since that time it has grown in importance at a steaily increasing rate and has become diversified into special fields such as immunohistochemistry, immunogenetic, and immunopathology. In the last 40 years, there has been a remarkable transformation in our understanding of the immune system and its function, largely due to advances in cell culture techniques, recombinant DNA methodology, and protein biochemistry.

The immune system is comparable in the complexity of its functions to the nervous system. Both systems are diffuse organs that are dispersed through most of the tissue of the body. Moreover, most lymphoid tissues receive direct sympathetic innervations to the blood vessels passing through the tissues and directly to lymphocytes, and lymphocytes express receptors for many neurotransmitters and neuropetides.

Bacteriologist Emil Adolf von Behring (1854–1917) (Fig. 1) is considered the father of immunology. He was awarded the first Nobel Prize for Physiology or Medicine in 1901 in recognition of his collaborative work with Japanese researcher Shibasaburo Kitasato (1856–1931) (Fig. 2) on serum therapy. These investigators studied the passive immune response to tetanus by injecting a healthy animal with serum from a diseased animal. The active components of the serum were called antitoxins because they neutralized the pathologic effects of bacterial toxins. They furthered their research by applying the same technique to the study and prevention of diphtheria. In 1890, von Behring and Kitasato observed symptoms in a horse that had been immunized with diphtheria and tetanus toxins. Von Behering described this reaction as hypersensitivity, a "paradoxical reaction." In 1893, he observed a reaction to diphtheria toxin in guinea pigs, and to antitoxin

Figure 1 A portrait of Emil Adolf Behring (Nobelprize.org).

Figure 2 A portrait of Shibasuro Kitasato (Nobelprize.org).

serum in man and attributed these ractions to the direct cumulative effect of the toxin rather than to an immune response.

In the early 1900s, Karl Landsteiner (1868–1943) showed that not only toxins, but also other, nonmicrobial substances, could induce humoral immunity. From these studies arose the more general term antibodies for the serum proteins that mediate humoral immunity and substances that bound antibodies were then called antigens.

The term "allergy" indicated that a patient has changed reactivity as a result of an antigenic stimulus. In man, allergic diseases might well include so many clinical conditions that in 1922 Arthur F. Coca introduced the word atopy meaning "strange disease." Coca defined atopy as "type of hypersensitiveness peculiar to man, subjected to hereditary influence, presenting the characteristic immediate vhealing type reaction, having circulating antibody reagin, and manifesting peculiar clinical syndromes such as asthma and hay fever" (Coca, 1922).

Lower animal forms possess so-called innate or nonspecific immune mechanisms, such as phagocytosis of bacteria by specialized cells, which afford them protection against infecting organism. Higher animals have evolved an adaptive or acquired immune response which provides a specific and more effective reaction to different infections. Specificity was mentioned as a fundamental feature of the adaptive immune response. The adaptive immune responses depend on the recognition of antigen by lymphocytes, a cell type that has evolved relatively recently, lymphocytes are present in all vertebrates, but not invertebrates. In any immune response the antigen induces clonal expansion in specific T and/or B cells, leaving behind a population of memory cells. These enable the next encounter with the same antigen to induce a secondary response, which is more rapid and effective than the normal primary response. The adaptive immunity provides specific protection against infection with bacteria, viruses, parasites, and fungi. It is able to provide rapid protection against a repeated challenge with the same or a similar foreign organism or toxin.

For the past century, immunology has fascinated and inspired some of the greatest scientists of our time and numerous Nobel prizes have been awarded for fundamental discoveries in immunology, from Paul Ehrlich's work on antibodies (1908) to the studies of Doherty and Zinkernagel (1974) elucidating mechanisms of cell-mediated immunity.

Figure 3 A portrait of Ilya Metchnikoff (Nobelprize.org).

The idea of cells directly involved in the defense of the body was first suggested by the zoologist Ilya Metchnikoff (1845–1916) (Fig. 3) in 1884. Based upon purely Darwinian evolutionary principles, Metchnikoff proposed in his famous book entitled "Immunity in the Infective Diseases" published in 1905 that phagocytic cell is the primary element in natural immunity (the first line of defense against infection) and is critical also for acquired immunity. Another notable contribution of the phagocytic theory was to the field of general pathology. In fact, at the time, most believed that inflammation was a damaging component of the disease itself; Metchinikoff, on the other hand, suggested that the inflammatory response was an evolutionary mechanism designed to protect the organism. As the developmental biologist, Scott Gilbert pointed out: "Metchnikoff was a comparative embryologist, and a very good one. One of the first comparative embryologist to study inverterbrates, Metchnikoff was ideally situated to discuss the origins of metazoans. (...) Metchnikoff entered into immunology through his attempt to prove that embryonic mesodermal cells had an intrinsic capacity for phagocytosis and that earliest metazoans, like the earliest embryonic stages, had a solely intracellular mode of digestion. This digestion was accomplished, Metchnikoff asserted, by the ameboid cells of the mesoderm. He would later frame the hypothesis that this primitive digestive function became a property of specialized phagocytes (i.e., macrophages) that would engulf and digest foreign objects such as pathogenic bacteria. This notion that the intracellular digestion found in protists would eventually give rise to the properties of immunocompetent cells is still a basic concept in modern immunology. Throughout his scientific career, Metchnikoff productively linked digestion, immunology, and evolution" (Gilbert, 1991).

In 1888, Louis Pasteur invited Metchnikoff to join him at the newly constructed "Pasteur Institute" in Paris, where Metchnikoff spent the next decades working productively to verify and extend the cellular theory of immunity. In 1908 the Swedish Academy conferred the Nobel Prize in Medicine jointly to Mechnikoff, the leading exponent of cellular theory, and to Paul Ehrlich (1854–1915) (Fig. 4), the leading exponent of humoral theory, "in recognition for their work on immunity."

Ehrlich identified for the first time mast cells when was a medical student. In his doctoral thesis, which was an admirable piece of forerunning insight, Ehrlich developed basic concepts about the microscopical features, histochemical behavior, and functional properties of these cells. Such studies led Ehrlich to formulate the concept of molecules that specifically bind to cell receptors and this principle was at the basis of the formulation of the side-chain theory of antibody formation. Then Ehrlich, decided to revisit small molecules with the aim of finding a "magic bullet" to kill microbial pathogens. During this work with dyes, Ehrlich tested the effects of methylene blue on malaria plasmodia. In 1909, Ehrlich discovered the first effective cure for syphilis, the "compound 606" (also called Salvarsan). Salvarsan was first tried on rabbits that had been infected with syphilis and then on patients with the dementia associated with the final stages of the disease. Salvarsan was used in the treatment of syphilis during the first half of the last century until it was superseded by penicillin. For this insight and this achievement, Ehrlich is known as the founder of chemotherapy.

Figure 4 A portrait of Paul Ehrlich.

In 1896, Ehrlich elaborated his side-chain theory to explain the appearance of antibodies in the circulation. He suggested that cells capable of forming antibodies possessed on their surface membranes specific side chains, which were receptors for antigens. He proposed that binding of antigen to the side chains provoked new synthesis of these side chains, which were liberated into serum as antibodies.

The cellular aspects of the same problem of antibody formation, as envisaged by Niels Jerne (1911–1994) and Sir Frank Macfarlane Burnet (1899–1985), are encompassed in the clonal selection theory. Burnet, Director of the "Walter and Eliza Hall Institute" in Melbourne, Australia, between 1957 and 1959 focused his effort on the elaboration of a theory of antibody synthesis, the clonal selection theory, which revolutioned theoretical immunology. The theory was capable of bridging the gap between physiological findings, such as the kinetics of antibody production, self-tolerance and immunological memory on the one hand, and the newest ideas on synthesis of proteins, on the other. Until about 1965, Burnet's clonal selection theory and the particular mechanism of variation he proposed (somatic mutation) were often considered inseparable. The three fundamental postulates of the clonal hypothesis are as follows: (1) the antigen receptor site is identical to the antigen combining site; (2) the specific link between the antigen receptor site and the synthesis of corresponding immunoglobulin is mere physical coexistence in the same cell; individual cells, that is, are specialized for the synthesis of a single species of antibody which serves as both antigen receptor site and antigen combining site; (3) the cell specialization postulated in (2) is inherited and therefore clonal. Moreover, according to clonal selection theory, antigens encountered after birth activate specific clones of lymphocytes, whereas when antigens are encountered before birth the result is the deletion of the clones specific for them, which Burnet termed "forbidden clones."

Brenner and Milstein (1966) proposed a somatic mutation theory in which hypervariability is caused by hypermutation. An enzyme repeatedly degrades one strand of the antibody gene and during its repair errors are introduced. Other somatic mutation theories hypothesized that hypervariation resulted from selection for mutation themselves occurring at a more or less normal rate. Jerne (1970) proposed that "nature exploits the fact that the most powerful selection pressure favoring mutants is the suppression of non mutants," and that an immune network existed within the body that interacted by means of idiotype recognition. In this context, when an antibody response is induced by an antigen, the antibody will in turn evoke an anti-idiotypic response to itself.

The modern era of antibody formation began in 1959 with investigations conducted at the Rockefeller University by Gerald Edelman and his colleagues, particularly M.D. Poulick. In 1961, Edelman showed that an immunoglobulin is made of two light and two heavy polypeptide chains (Edelman and Poulik, 1961). In 1965 the first sequence of a light chain was published and it resulted made of a constant region and a variable region (Hilschmann and Craig, 1965). In 1969, Edelman published the complete sequence of an immunoglobulin showing that both the heavy and the light chains are made of constant and variable regions, that the variable regions contribute to the antigen binding site and that the tertiary structure of the antibody is based on repeated globular regions that he called "domains" (Edelman, 1973).

Mammalian stem cells differentiate into several kinds of blood cell within the bone marrow. During this process (hematopoiesis), all lymphocytes originate from a common lymphoid progenitor before differentiating into their distinct lymphocyte types. B cells mature into B lymphocytes in the bone marrow, while T cells migrate to, and mature in, a distinct organ called the thymus. Following maturation the lymphocytes enter the circulation and peripheral lymphoid organs where they survey for invading pathogens and/or tumor cells. In birds the lymphoid tissue is considered to consist of two vertical compartments of differentiating lymphocytes: one responsible for the discrimination of self and nonself and the expression of cell-mediated immunity, and the other responsible for the production of immunoglobulins and antibodies. The former in birds and in mammals, is dependent upon the thymus (T-dependent system) and the latter, in birds, upon the cloacal bursa (B-dependent system).

The bursa of Fabricius is a lymphoepithelial organ located near the cloaca. Just as the thymus appears to act as a central lymphoid organ controlling the maturation of lymphocytes concerned largely with cell-mediated immunity, so the bursa of Fabricius is responsible for the development of immunocompetence in cells destined to make humoral immunity. In 1956, for the first time Bruce Glick and Timothy Chang reported that the bursa of Fabricius plays an important role in the antibody production. Their demonstration that antibody responses are suppressed in the majority of bursectomized chickens become the cornerstone of modern immunology. After the discovery of Glick, it has been demonstrated that the mammalian equivalent of the bursa of Fabricius is the bone marrow. The development of an antibody response is a culmination of a series of cellular and molecular interactions occurring in an orderly sequence between a B cell and a variety of other cells of the immune system.

The thymus is one of the two primary lymphoid organs. It is responsible for the provision of T-lymphocytes to the entire body and provides a unique microenvironment in which T-cell precursors (thymocytes) undergo development, differentiation, and clonal expansion. The Australian scientist Francis Albert Pierre Miller gave a fundamental contribution to the description for the first time of the crucial role of the thymus for normal development of the immune system. Robert Alan Good, a pioneer in the field of immunodeficiency diseases, defined the cellular basis and functional consequences of many of the inherited immunodeficiency diseases. He contributed to the discovery of the pivotal role of the thymus in the immune system development and defined the separate development of the thymus-dependent and bursa-dependent lymphoid cell lineages and their responsibilities in cell-mediated and humoral immunity.

Immunodeficiency diseases result from the absence of failure of normal functions of one or more elements of the immune system and cause increased susceptibility to infection in patients. Congenital abnormalities in B cell development and function results in deficient antibody production. Clinically the disorders are characterized by recurrent infections with pyogenic organisms, such as pneumococcus, *Hemophilus inflenzae*, and streptococcus. The congenital absence of the thymus, as occurs in Di George syndrome in humans or in nude mouse strain, is characterized by low numbers of mature T cells in the circulation and peripheral lymphoid tissues and functional deficiencies in T-cell-mediated immunity.

REFERENCES

Brenner, S., Milstein, C., 1966. Origin of antibody variation. Nature 211, 242.

Coca, A.F., 1922. Studies in specific hypersensitiveness: the preparation of fluid extracts and solution for use in the diagnosis and treatment of the allergies with notes on the collection of pollens. J. Immunol. 7, 163.

Doherty, P.C., Zinkernagel, R.M., 1974. T-cell-mediated immunopathology in viral infections. Transplant. Rev. 19, 89–120.

Edelman, G.M., 1973. Antibody structure and molecular immunology. Science 180, 830–839.

Edelman, G.M., Poulik, M.D., 1961. Studies on structural units of the gamma-globulins. J. Exp. Med. 113, 861–884.

Ehrlich, P., 1905. Collected Studies in Immunity. John Wiley & Sons, New York.

Gilbert, S., 1991. Foreword. In: Tauber, A.L., Chernyak, L. (Eds.), Metchnikoff and the Origins of Immunology. From Metaphor to Theory. Oxford University Press, New York-Oxford.

Hilschmann, N., Craig, L.C., 1965. Amino acid sequence studied with Bence Jones proteins. Proc. Natl. Acad. Sci. U.S.A. 53, 1403–1409.

Jerne, N.K., 1970. The somatic generation of immune recognition. Eur. J. Immunol. 1, 1.

Metchnikoff, E., 1905. Immunity in the Infectious Diseases. Macmillan, New York.

Talmage, D.W., 1986. The acceptance and rejection of immunological concepts. Ann. Rev. Immunol. 4, 1–11.

FURTHER READING

Burnet, F.M., 1959. The clonal selection theory of acquired immunity. Cambridge University Press, Cambridge.

Metchnikoff, E., 1893. Lectures on the Comparative Pathology and Inflammation. Kegan, Paul, Trench, Trübner, London.

Silverstein, A.M., 1988. A History of Immunology. Academic Press, New York.

Tauber, A.I., Chernyak, L., 1991. Metchnikoff and the Origins of Immunology. Oxford University Press, New York.

CHAPTER 1

The Contribution of Bruce Glick to the Definition of the Role Played by the Bursa of Fabricius in the Development of the B Cell Lineage

1.1 FABRICIUS AB AQUAPENDENTE AND THE DISCOVERY OF THE BURSA

Student and successor of Andreas Vesalius (1514–64) and Gabriel Fallopius (1523–62), Girolamo Fabrici or Fabrizio (Hieronymus Fabricius ab Aquapendente) (1533–1619) was Professor of Surgery from 1565 to 1613 and practiced and taught anatomy at the University of Padova (Fig. 1.1) (Smith et al., 2004). In 1594, he built the first permanent theater ever designed for public anatomy dissection. William Harvey was one of his pupils. In addition to his demonstration of the valves of the veins, Fabricius is best known for his description of the bursa that bears his name. A manuscript entitles "*De Formatione Ovi et Pulli*," found among his lecture notes was published in 1621, contains the first description of the bursa (Adelman, 1967): "The third thing which should be noted in the podex is the double sac [bursa] which in its lower portion projects toward the pubic bone and appears visible to the observer as soon as the uterus already mentioned presents itself to view."

The sac-like organ has ever since been known as the bursa of Fabricius. It overlies the dorsal surface of the terminal portion of the gut in birds (Fig. 1.2). In the 5/6-day chick embryo, it arises as a dorsocaudal outpouching near the cloaca that takes the form of a median lamina of endodermal epithelium permeated with spherical vacuoles of various sizes that eventually coalesce to create a lumen (Hamilton, 1952). The bursa grows considerably during development and changes from round to oval, and hypertrophy of the mesoderm surrounding the bursal epithelium produces longitudinal plicae that project into its lumen (Romanoff, 1960). The bursa, at first only epithelial, is invaded by stem cells of yolk sac or fetal liver origin undergoing rapid proliferation and reaches its maximum size at 8–10 weeks of age (Ciriaco et al., 2003).

Milestones in Immunology. DOI: http://dx.doi.org/10.1016/B978-0-12-811313-4.00001-2

Figure 1.1 A portrait of Fabricius ab Aquapendente.

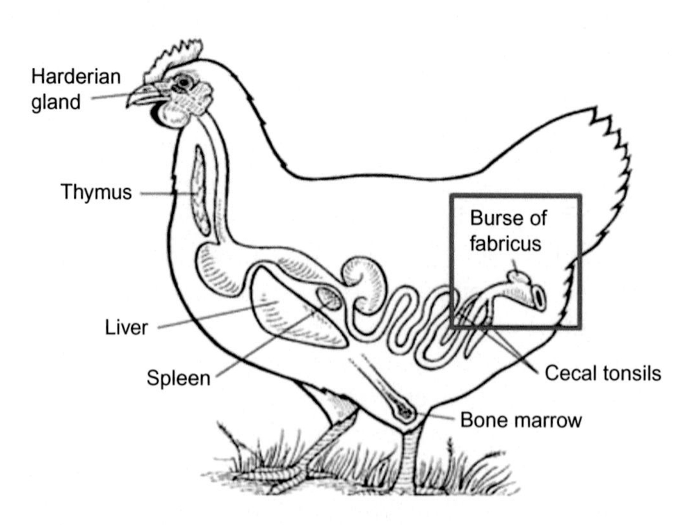

Figure 1.2 A drawing showing the anatomical position of the bursa of Fabricius.

Between the 13th and 15th day, epithelial cells lining the plicae thicken and extend into the tunica propria as epithelial buds, which then separate from the epithelium. Lymphopoiesis is active in cells that form the medullis of the bursal follicle (Ackerman and Knouff, 1959; Ackerman, 1962).

Follicles may be present during late embryonic development (after 16 days), and they are best observed by light microscopy at hatching and during the early growth of the bursa of Fabricius (Frazier, 1974). The bursa of Fabricius has 8000 to 12,000 total follicles (each of which contains 1000 bursal cells), each composed of a cortex, medulla, corticomedullary border, and follicle-associated epithelium (FAE) (Glick, 1983). Lymphopoiesis is active in the medulla of follicles (Ackerman and Knouff, 1959).

Medullary anlage emerges on 11–12 day of incubation followed by the formation of the FAE on 14–15 day (Bockman and Cooper, 1973). Medulla consists of epithelial cells and blood-borne hematopoietic cells, including dendritic cells, lymphoid cells, and macrophages. The bursa provides a unique microenvironment essential for the proliferation and differentiation of B cells (Ratcliffe, 2006). B cell progenitors colonize the bursa from 8 to 14 incubation day (Le Douarin et al., 1975). Medullary B cells express surface immunoglobulin M (IgM), and the first surface IgM-positive cells are detected from 12 incubation day and at hatching more than 90% of bursal cells are mature B cells. Cells that fail to express surface antibodies are eliminated by apoptosis. Only B cell precursors that positively rearrange the immunoglobulin gene express surface immunoglobulin and expand in bursal follicles (Ratcliffe and Jacobsen, 1994). The presence of M cells within this epithelium explains the movement of antigen from the lumen into the medulla, where immature B cells develop (Sayesh et al., 2000).

1.2 THE BURSA OF FABRICIUS PLAYS A MAJOR ROLE IN THE DEVELOPMENT OF ANTIBODY-MEDIATED IMMUNITY

In December 1952, Bruce Glick (Fig. 1.3) at the Ohio State University demonstrated that the bursa of Fabricius grow most rapidly during the first 3 weeks after hatching. He thus became convinced that functional investigation of the bursa would only be successful if it was removed (bursectomy) within this period.

Figure 1.3 A portrait of Bruce Glick.

In 1954, Timothy S. Chang, a graduate student of Glick, needed birds to develop antibody against *Salmonella*. The only ones available were those of Glick. He therefore injected 6-month-old pullets with *Salmonella*-type O antigen to obtain serum with a high antibody titer. Several pullets died subsequently the immunization, and none of those that survived produced antibody. It was then found that the entire batch had been bursectomized during the period of rapid bursa growth.

Glick (1955) deduced that absence of bursa of Fabricius was responsible for this failure, since nonbursectomized pullets produced normal antibody titers and designed two experiments to substantiate this conclusion. Equal numbers of male and female White Leghorns were bursectmized at 12 day of age and injected six times at intervals of 4 days with *Salmonella typhimurium* O antigen. At 7 weeks, 7/10 bursectomized birds and 2/10 controls failed to produce antibody (Glick, 1955). The second experiment employed larger numbers of birds and two breeds: 89.3% of the bursectomized birds failed to produce antibody compared with only 13.7% of the controls (Chang et al., 1955; Glick et al., 1956).

1.3 BURSECTOMY DO NOT ABROGATE THE ANTIBODY RESPONSE TO CELLULAR ANTIGENS

Next, Chang et al. (1957) showed that bursectomy at 2 weeks was more effective in suppressing antibody production than at 5 or 10 weeks of age. Failure of bursectomy to eliminate all antibody production suggested the existence of a brief period in embryo development during which the bursa of Fabricius could be functional.

The first experiments to evaluate the existence of a functional period for the bursa of Fabricius (Meyer et al., 1959) took advantage of the regressive influence of androgens on the posthatched bursa of Fabricius (Kirkpatrick and Andrews, 1944; Glick, 1957). Treatment of 9 to 12-day embryo with testosterone prevented immunoglobulins production and lymphoid development and, presumably, destroyed the stem cells which are necessary for B cells production.

Subsequent injection of bovine serum albumin (BSA) into chicks hatched from eggs injected with testosterone on day 5 of incubation revealed complete immunoglobulins elimination, whereas chicks from eggs injected with testosterone on day 12 or 13 possessed significantly reduced levels of antibody (Mueller et al., 1960, 1962). The bursa of Fabricius was generally absent in 19-day embryos that had received testosterone prior to the 8 day of incubation (Warner and Burnet, 1961). Hormonal bursectomy enhanced graft versus activity of injected homologous cells and allowed allogenic spleen cells to be more effective in antibody synthesis (Papermaster et al., 1962).

Various bursectomy methods cause more or less complete B cells defect and agammaglobulinemia. They include testosterone treatment (Meyer et al., 1959; Glick, 1957, 1964; Glick and Sadler, 1961; Warner and Burnet, 1961; Papermaster et al., 1962), cyclophosphamide administration (Lerman and Weidanz, 1970; Eskola and Toivanen, 1974), colchicine treatment (Romppanen and Sorvari, 1980), x-irradiation (Cooper et al., 1966), and surgical operations (Van Alten et al., 1968; Fitzsimmons et al., 1973).

Cooper et al. (1966) showed that chickens irradiated at hatching and also subjected to total bursectomy develop peripheral small lymphocytes in a normal fashion, reject skin syngeneic grafts and display normal graft versus host reactions. They are, however, prevented from developing the two clearly definable immunoglobulins and are completely unable to form circulating antibodies, even in response to strong antigenic stimulation.

1.4 IMMUNOGLOBULIN SYNTHESIS REGULATION

The bursa of Fabricius as a site of antibody synthesis was investigated by Glick and his coworkers in the early 1960s. Two experiments gave conflicting results. In the first, pheasant bursa cells produced antibody to bovine immunoglobulins (Kerstetter et al., 1962), whereas in the second, the bursa of Fabricius was unable to produce plaque-forming cells to sheep red blood cells (SRBC) (Dent and Good, 1965). The reason why pure B cells did not produce antibodies against SRBC only became obvious later when the group of Henry Claman discovered the requirement of B–T-cell cooperation for antibody production against this and other T-dependent antigen (Claman and Chaperon, 1969). These authors investigated the participation of both T and B cells in the in vitro response of spleen cells from mice immunized with the hapten NIP coupled to a nonimmunogenic isologous gamma globulin carrier (mouse gamma globulin (MGG)) (Claman and Chaperon, 1969).

Glick failed to identify antibody to bovine serum albumin (BSA) in the bursa of Fabricius from 3-week-old intravenously immunized chickens (Glick and Whatley, 1967). The B cell differentiates in the bursa of Fabricius and is able to produce immunoglobulins on the 14th day of embryo development. The first immunoglobulin is the large 1,000,000 molecular weight molecule called IgM, followed by immunglobulin G (IgG) on the 20th and then immunoglobulin A (IgA) (Cooper et al., 1969; Kincade and Cooper, 1971). Two equally plausible explanations of this sequence were advanced. One held that IgM B cells give to the IgG and IgA B cells, the other proposed sequential intrabursal development of isotype-committed sublineages. Kincade and Cooper (1973) found that the anti-μ mediated inhibition of IgM B cells also inhibited development of the IgG and IgA B cells. Moreover, the combination of embryonic anti-μ administration and posthatching bursectomy resulted in permanent agammaglobulinemia. These experiments indicated that while all chicken B cells express IgM initially, they can switch to the production of other isotypes. Neonatal anti-μ antibody treatment also inhibited mouse-B cell development and antibody production of all Ig isotypes (Lawton et al., 1972).

1.5 DELINEATION OF THE THYMIC AND BURSAL LYMPHOID SYSTEMS IN THE CHICKEN

As Cooper (2015) pointed out: "Chickens offered an animal in which to test the possibility of alternative lymphocyte lineage, although it was unclear at the time whether the thymus and the bursa had synergistic or independent roles and just how they might function. It proved difficult to show the early thymectomy affected either cellular or humoral immunity, probably because of the fairly mature status of the immune system in newly hatched chicks. Defining the respective roles of the thymus and the bursa would thus require either removing one or the other early in embryonic life or removing them after hatching in conjunction with the destruction of cells that had developed earlier under their influence."

Functional dissociation of the chicken immune system based on differences in thymic and bursal influences was originally suggested by Szenberg and Warner (1962). Following the Glick' demonstration of the crucial function of the bursa of Fabricius in development of antibodies and the immune responses related to their production, in 1958 Francis A. P. Miller in Australia discovered the role of thymus-derived cells for cellular immunity (Ribatti et al., 2006). Miller's experiments indicate that thymectomy is associated generally with a diminution in the lymphocyte population and the earlier in life thymectomy is performed, the greater the deficiency of lymphocytes in other lymphoid organs (Ribatti et al., 2006). Robert Good and his collaborators (notably Max D. Cooper) developed the idea of the B- and T-cell concepts, demonstrating the essential role of the thymus in development of cellular immunity functions other than antibody production in the chickens (Cooper et al., 1965, 1966). Chickens were thus the first source of the two-component concept of immunity. Subletal x-irradiation of newly hatched chickens was needed to clarify the roles of the thymus and the bursa of Fabricius in development of the two separate and functionally different lymphoid systems (Cooper et al., 1965). The bursectomized and irradiated birds were completely devoid of germinal centers, plasma cells and the ability to produce antibodies, yet they had perfectly normal development of thymocytes and lymphocytes elsewhere in the body that mediated cellular immune reactions, while the thymectomized and irradiated birds were deficient in lymphocytes that mediated cellular immunity as assessed by skin graft rejection, delayed-type hypersensivity and graft-vs-host reaction, but still produced germinal centers, plasma cells and circulating immunoglobulins. Van Alten et al. (1968) used bursectomy within the eggs to show that the two-component concept was clearly evident even in the absence of x-irradiation. The bursa of Fabricius and the thymus are "central lymphoid organs" in the chicken, essential for the ontogenetic development of their adaptive immunity. Surgical removal of one or both of organs in the newly hatched chickens, followed by sublethal x-irradiation, led to the recognition of two morphological distinct cell systems in the peripheral lymphoid tissues of the spleen, gut and other organs, and clear differentiation of their functions. The thymus controls development of all cell-mediates immunities, including delayed reactions, allograft immunities, and other immunologic functions. Good in addition to being a basic immunologist also held the position of Professor of Pediatrics at the School of Medicine in Minneapolis, Minnesota. He thus had access to the various cases of immunodeficiencies that had led him to recognize similarities between Bruton's agammaglobulinemia and Glick's bursectomy on one hand and Di George syndrome and Miller's thymectomized mice on the other hand.

In fact, removal of the bursa of Fabricius from the egg inhibits germinal centers, plasma cells and prevented antibody production (Perey and Good, 1968). Bursectomized chickens are strikingly like patients with Bruton's x-linked agammaglobulinemia (Peterson et al., 1965), and in ovo thymectomized chicks are strikingly those with Di George syndrome, while with severe combined immunodeficiency disease (SCID) are like chickens bursectomized and thymectomized in the newly hatched period (Peterson et al., 1965). The major immunodeficiencies, Bruton's disease, Di George syndrome, and SCID, are thus mimicked by bursectomized or thymectomy in ovo.

1.6 BURSA OF FABRICIUS EQUIVALENT IN MAMMALS AND OTHER VERTEBRATES

As Miller (2002) pointed out: "As early in 1962, Burnet raised urgent question of whether there is a functional equivalent of the bursa in the mammals (Burnet, 1962a)." My work had, however, shown that neonatal thymectomy in the mouse was associated not only with defective cellular immunity but also with antibody impaired-producing capacity to certain antigens, which later known as thymus-dependent antigens. Probably, this work led Burnet to the view that in "mammals it is highly probable that the thymus also carries out the function performed by the bursa of Fabricius in the chicken, which is to feed into the body the cells whose descendants will produce antibodies (Burnet, 1962b)."

The bursa of Fabricius is present in all avian orders, but is absent in mammals. Several structures, however, have been identified as "bursa equivalents," such as gut-associated lymphoid tissues in rabbits and ungulates and the bone marrow in rodents and primates, including humans. Archer et al. (1963) found that the rabbit sacculus rotundus located at the ileo-coecal valve, like the bursa of Fabricius, develop within follicular outpouchings of the lower gut. Immediate extirpation of this organ in neonate resulted in an impressive and lifelong immunodeficiency of antibody production (Perey and Good, 1968; Archer et al., 1963). Knight and Crane (1994) have since demonstrated that the bursa of Fabricius and the appendix-sacculus rotundus mediate very similar influences on the humoral system. However, the sacculus rotundus has not emerged as the bursa of Fabricius equivalent organ. Owen et al. (1974) found that Ig-bearing cells first appear in the liver during mouse embryogenesis and employed fetal liver organ cultures to show that B cells are generated in this hematopoietic tissue. Moreover, Owen et al. (1975) found that, after their colonization with hematopoietic stem cells, fetal long bones can also generate B cell *ex vivo*. This finding suggested that mammalian B cell generation is a multifocal process that shifts from one hematopoietic environment to another during development, to continue throughout life in the bone marrow. It is now clear that in mammals B cells stay and differentiate in the bone marrow, a very convenient ethymological coincidence, since the nomenclature for B cells as bone marrow derived cells had not be changed.

Different structures have been identified as bursal-equivalent in mammals, including gut-associated lympho-epithelial tissues (GALTs), bronchus-associated lymphoid tissues, and the bone marrow in primates, including humans. The GALT comprises lymphoid cells residing in epithelial lining and distributed in the underlying lamina propria as well as specialized lymphoid structures, including Meckel's diverticulum, Peyer's patches, and coecal tonsils. Neonatal appendectomy followed by Peyer's patch removal in combination with whole body irradiation to destroy preexisting lymphocytes in rabbits induced immunological defects comparable to those observed in older chickens subjected to bursectomy and irradiation (Cooper et al., 1966).

REFERENCES

Ackerman, G.A., 1962. Electron microscopy of the bursa of Fabricius of the embryonic chick with particular reference to the lympho-epithelial nodules. J. Cell. Biol. 13, 127–146.

Ackerman, G.A., Knouff, R.A., 1959. Lymphocytopoiesis in the bursa of Fabricius. Am. J. Anat. 104, 163–205.

Adelman, H.B., 1967. The Embryological Treatites of Hieronymus Fabricius of Aquapendente, 1. Cornell Univ. Press, Ithaca, New York, pp. 147–191.

Archer, O.K., Sutherland, D.E.R., Good, R.A., 1963. Appendix of the rabbit: a homologous of the bursa in the chicken? Nature 200, 337–339.

Bockman, D.E., Cooper, M., 1973. Pinocytosis by epithelium associated with lymphoid follicles in the bursa of Fabricius, appendix, and Peyer's patches. An electron microscopic study. Am. J. Anat. 136, 455–478.

Burnet, F.M., 1962a. The role of thymus and related organs in immunity. Br. Med. J. 2, 807–811.

Burnet, F.M., 1962b. The thymus gland. Sci. Am. 207, 50–57.

Chang, T.S., Glick, B., Winter, A.R., 1955. The significance of the bursa of Fabricius of chickens in antibody production. Poult. Sci. 34, 1187.

Chang, T.S., Rheins, M.S., Winter, A.R., 1957. The significance of the bursa of Fabricius in antibody production in chickens. Poult. Sci. 36, 735–738.

Ciriaco, E., Pinera, P.P.P., Diaz Esnol, B., et al., 2003. Age-related changes in the avian primary lymphoid organs (thymus and bursa of Fabricius). Microsc. Res. Tech. 62, 482–487.

Claman, H.N., Chaperon, E.A., 1969. Immunologic complementation between thymus and marrow cells. A model for the two-cell theory of immunocompetence. Transplant. Rev. 1, 92–113.

Cooper, M.D., 2015. The early history of B cells. Nat. Rev. Immunol. 15, 191–197.

Cooper, M.D., Peterson, R.D.A., Good, R.A., 1965. Delineation of the thymic and bursal lymphoid systems in the chicken. Nature 205, 143–146.

Cooper, M.D., Peterson, R.D.A., South, M.A., et al., 1966. The functions of the thymus system and the bursa system in the chicken. J. Exp. Med. 123, 75–102.

Cooper, M.D., Cain, W.A., Van Alten, P.J., et al., 1969. Development and function of the immunoglobulin producing system. I. Effect of bursectomy at different stages of development on germinal centers, plasma cells, immunoglobulins and antibody production. Int. Arch. Allergy Appl. Immunol. 35, 242–252.

Dent, P.D., Good, R.A., 1965. Absence of antibody production in the bursa of Fabricius. Nature 207, 491–493.

Eskola, J., Toivanen, P., 1974. Effect of in ovo treatment with cyclophosphamide on the lymphoid system in chicken. Cell. Immunol. 13, 459–471.

Fitzsimmons, R.C., Garrod, E., Garnett, I., 1973. Immunological responses following early embryonic surgical bursectomy. Cell. Immunol. 9, 377–383.

Frazier, J.A., 1974. The ultrastructure of the lymphoid follicles of the chick bursa of Fabricius. Acta Anat. 88, 385–397.

Glick B, 1955, Ph.D. Dissertation. Ohio State University, Columbus. pp. 1–102.

Glick, B., 1957. Experimental modification of the growth of the bursa of Fabricius. Poult. Sci. 36, 18–23.

Glick, B., 1964. The bursa of Fabricius and the development of immunologic competence. In: Good, R.A., Gabrielsen, A.E. (Eds.), The Thymus Immunobiology. Hoeber, New York, pp. 345–358.

Glick, B., 1983. Bursa of Fabricius. Avian Biol. 7, 443–450.

Glick, B., Sadler, C.R., 1961. The elimination of the bursa of Fabricius: reduction of antibody production in birds from eggs dipped in hormone solutions. Poult. Sci. 40, 185–189.

Glick, B., Whatley, S., 1967. The presence of immunoglobulins in the bursa of Fabricius. Poult. Sci. 46, 1587–1589.

Glick, G., Chang, T.S., Jaap, R.G., 1956. The bursa of Fabricius and antibody production. Poult. Sci. 35, 224–234.

Hamilton, H.L., 1952. Lillie's Development of the Chick. Holt, Rinehart and Winston, New York, pp. 390–391.

Kerstetter Jr, T.H., Buss, I.O., Went, H.A., 1962. Antibody-producing function of the bursa of Fabricius of the ring-necked pheasant. J. Exp. Zool. 149, 233–237.

Kincade, P.W., Cooper, M.D., 1971. Development and distribution of immunoglobulin-containing cells in the chicken. An immunofluorescent analysis using purified antibodies to mu, gamma and light chains. J. Immunol. 106, 371–382.

Kincade, P.W., Cooper, M.D., 1973. Immunoglobulin A. Site and sequence of expression in developing chicks. Science 179, 398–400.

Kirkpatrick, C.M., Andrews, F.N., 1944. Body weights and organ measurements in relation to age and season in ring-necked pheasants. Endocrinology 34, 340–345.

Knight, K.L., Crane, M.A., 1994. Generating the antibody repertoire in rabbit. Adv. Immunol. 56, 179–218.

Lawton, A.R., Asofsky, R., Hylton, M.B., et al., 1972. Suppression of immunoglobulin class synthesis in mice. I. Effects of treatment with antibody to μ-chain. J. Exp. Med. 135, 277–297.

Le Douarin, N.M., Houssaint, E., Joterau, V., et al., 1975. Origin of hemopoietic stem cells in embryonic bursa of Fabricius and bone marrow studied by interspecific chimeras. Proc. Natl. Acad. Sci. U.S.A. 72, 2701–2705.

Lerman, S.P., Weidanz, W.P., 1970. The effect of cyclophosphamide on the ontogeny of the humoral immune response in chickens. J. Immunol. 105, 614–619.

Meyer, R.K., Rao, M.A., Aspinall, R.L., 1959. Inhibition of the development of the bursa of Fabricius in the embryos of the common fowl by 19-nortesterone. Endocrinology 64, 890–897.

Miller, J.F.A.P., 2002. The discovery of thymus function and of thymus-derived lymphocytes. Immunol. Rev. 185, 7–14.

Mueller, A.P., Wolfe, H.R., Meyer, R.K., 1960. Precipitin production in chickens. XXI. Antibody production in bursectomized chickens and in chickens injected with 19-nortestosterone on the fifth day of incubasion. J. Immunol. 85, 172–179.

Mueller, A.P., Wolfe, H.R., Meyer, R.K., et al., 1962. Further studies on the role of the bursa of Fabricius in antibody production. J. Immunol. 88, 354–360.

Owen, J.J.T., Cooper, M.D., Raff, M.C., 1974. In vitro generation of B lymphocytes in mouse foetal liver,-a mammalian "bursa equivalent". Nature 249, 361–363.

Owen, J.J.T., Raff, M.C., Cooper, M.D., 1975. Studies on the generation of B lymphocytes in the mouse embryo. Eur. J. Immunol. 5, 468–473.

Papermaster, B.W., Friedman, D.I., Good, R.A., 1962. Relationships of the bursa of Fabricius to immunologic responsiveness and homograft immunity in chicken. Proc. Soc. Exp. Biol. Med. 110, 62–64.

Perey, D.Y.E., Good, R.A., 1968. Experimental arrest and induction of lymphoid development in intestinal lymphoepithelial tissues of rabbits. Lab. Invest. 18, 15–26.

Peterson, R.D.A., Cooper, M.D., Good, R.A., 1965. The pathogenesis of immunologic deficiency diseases. Am. J. Med. 38, 579–604.

Ratcliffe, M.J.H., 2006. Antibodies, immunoglobulin genes and the bursa of Fabricius in chicken B cell development. Dev. Comp. Immunol. 30, 101–118.

Ratcliffe, M.J.H., Jacobsen, K.A., 1994. Rearrangement of immunoglobulin genes in chicken B cell development. Semin. Immunol. 6, 175–184.

Ribatti, D., Crivellato, E., Vacca, A., 2006. Miller's seminal studies on the role of thymus in immunity. Clin. Exp. Immunol. 144, 371–375.

Romanoff, A.L., 1960. The avian embryo. Macmillan, New York, pp. 497–508.

Romppanen, T., Sorvari, T.E., 1980. Chemical bursectomy of chickens with colchicine applied to the anal lips. Am. J. Pathol. 100, 193–208.

Sayesh, C.E., Demaries, S.L., Pike, K.A., et al., 2000. The chicken B-cell receptor complex and its role in avian B-cell development. Immunol. Rev. 175, 187–200.

Smith, S.B., Macchi, V., Parenti, A., De Caro, R., 2004. Hieronymus Fabricius ab Acquapendente (1533–1619). Clin. Anat. 17, 540–543.

Szenberg, A., Warner, N.L., 1962. Dissociation of immunological responsiveness in fowls with hormonally development of lymphoid tissues. Nature 194, 146.

Van Alten, P.J., Cain, W.A., Good, R.A., , et al.,Cooper, M.D. 1968. Gamma globulin production and antibody synthesis in chickens bursectomized as embryos. Nature 217, 358–360.

Warner, N.L., Burnet, F.M., 1961. The influence of testosterone treatment on the development of the bursa of Fabricius in the chick embryo. Aust. J. Biol. Soc. 14, 580–587.

CHAPTER 2

The Contribution of Robert A. Good and Francis A.P. Miller to the Discovery of the Role of Thymus in Immunity

2.1 BIOGRAPHIC NOTES

Robert A. Good (Fig. 2.1) was born on May 21, 1922, in Crosby, Minnesota. At the University of Minnesota, he was the first student to obtain MD and PhD degrees at the age of 25. He received training in Pediatrics at the University of Minnesota Hospital, followed by an immunological fellowship at the Rockfeller Institute. He began his intellectual and experimental queries related to the thymus in 1952 at the University of Minnesota, initially with pediatric patients. However, his interest in the plasma cell, antibodies, and the immune response began in 1944, while still in Medical School at the University of Minnesota in Minneapolis, with publications beginning in 1945 (Good and Campbell, 1945).

Dr. Good was professor of pediatrics at 32 years old and professor of microbiology and pathology and chaired the Department of Pathology in 1970. In 1973, he moved to New York City to become president of the Sloan-Kettering Institute for Cancer Research and professor of pediatrics, medicine, and microbiology–immunology. In 1985, he joined the University of South Florida, School of Medicine in Tampa and professor of pediatrics, medicine, and microbiology–immunology. He died in 2003. Throughout his career, Dr. Good has an exceptional record of more than 100 honors and awards, including 13 honorary doctorate degree and multiple lifetime achievement awards.

Jacques Francis A. P. Miller (Fig. 2.2) is Professor Emeritus at the Walter and Eliza Hall Institute of Medical Research (WEHI) and at the University of Melbourne. He was Head of Experimental Pathology Unit at the WEHI (1966–96) and is recognized as having discovered the function of the thymus. Miller and his PhD student Graham Mitchell proved the existence and function of T cells and B cells, which

Milestones in Immunology. DOI: http://dx.doi.org/10.1016/B978-0-12-811313-4.00002-4

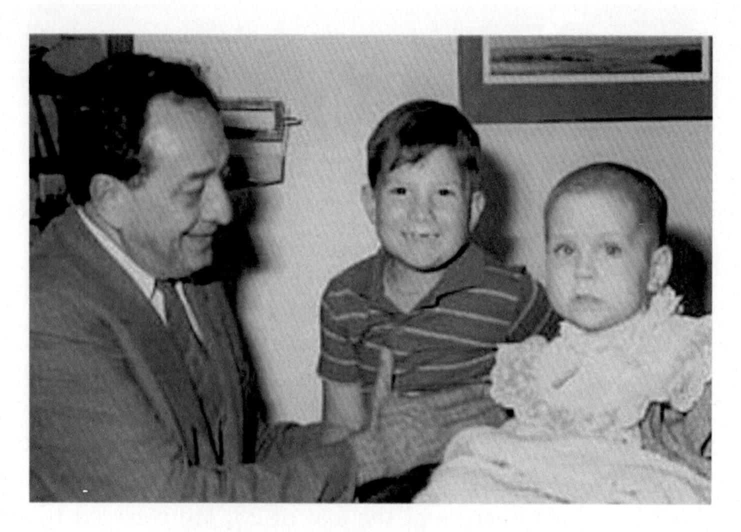

Figure 2.1 A portrait of Robert A. Good with two young patients.

Figure 2.2 A portrait of Francis Albert Pierre Miller.

has significantly opened up whole new fields for the study of immunology, including the study of cancer, autoimmune disease, transplantation, and human immunodeficiency virus (HIV) and acquired innune deficiency syndrome (AIDS). Jacques Miller continues to be one of the most respected research thymus biologists in the world (Crivellato et al., 2004) (Fig. 2.3).

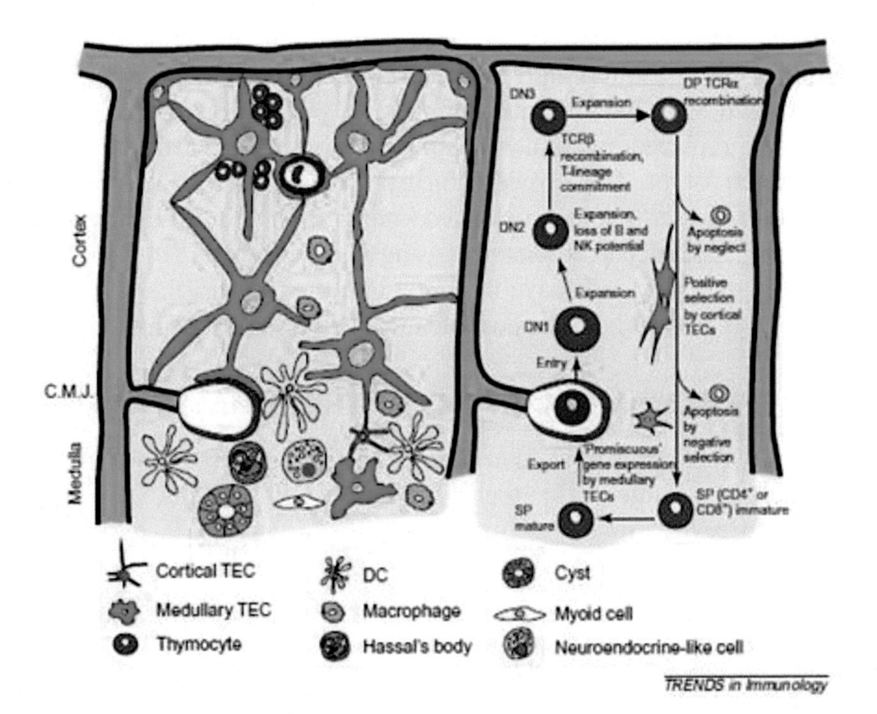

Figure 2.3 Structural and functional architecture of the thymus. Legend: TEC, *thymic epithelial cells;* DC, *dendritic cells;* TCR, *T-cell receptor;* DN, *double negative thymocytes;* DP, *double-positive thymocytes;* SP, *single-positive thymocytes.* Reproduced form Crivellato E, Vacca A, Ribatti D. Setting the stage: an anatomists's view of the immune system. Trends. Immunol. 2004; 25: 201–207.

2.2 THE THYMUS AND ITS FUNCTION

By the 1950s, recognition of the thymus as the site of production of lymphocytes had been well established. Their immunological competence was unequivocally demonstrated by Billingham et al. (1956) and by Gowans et al. (1962).

During the early stages of embryogenesis, lymphocytes were demonstrated to differentiate from the epithelial component of the thymus rudiment (Auerbach, 1961). Once formed within the embryonic thymus, they migrate out, colonize the spleen and lymph nodes, and constitute the immunologically competent cells of the lymphoid system (Auerbach, 1963).

Cells of bone marrow origin represent the precursors which in irradiated organism may differentiate to immunologically competent lymphocytes in response to an inductive stimulus of the thymus (Feldman and Globerson, 1964). Experiments in animals which were thymectomized, irradiated, "reconstituted" with bone marrow cells bearing a chromosomal marker and grafted with an unmarked thymus have shown that the original lymphocyte population of the grafted thymus is replaced by a new population of cells bearing the bone marrow karyotype; this provides further evidence that the lymphocyte population of the thymus arises from immigration and differentiation of blood-borne stem cell precursors, arising in this instance from the bone marrow. Once stem cells have migrated into the thymus, they differentiate into thymic lymphocytes, possible under local inductive influences.

2.3 THE FUNCTIONAL ANATOMY OF THE HUMAN THYMUS

Human thymus receives stem cells from the bone marrow and provides the microenvironment for them to develop in T cells, which are released to begin a long-life circulating and recirculating through blood and lymph, slowly moving through T-cell zones in peripheral lymphatic tissue. The thymus is the first lymphoid organ to develop followed in turn by the central lymph nodes, spleen, peripheral lymph nodes, and gut. The thymus and the parathyroid glands develop from epithelial anlagen of the third and, to a lesser extent, from the fourth pharyngeal pouches. It develops from an ectodermal–endodermal juncture, and its epithelial components contain derivative of both ectodermal and endodermal germ layers. The thymic mass gradually increases with colonization of blood-borne hematopoietic stem cells (Le Douarin and Jotereau, 1975), and the rapid increase, a few days before hatching, results in the appearance of the medulla. Uncommitted hematopoietic progenitors, therefore, enter the land through postcapillary venules at the cortico-medullary junction. Hematopoietic stem cells invade the epithelial anlage, and they first move toward the subcapsular region and acquire T lineage commitment.

During its development, the thymus undergoes a descensus which brings it to lie in the anterior mediastinum, in close connection with the pericardium and the great veins at the base of the heart where the endodermal epithelial masses fuse in the midline in the 12th week of embryonic life. The lower border of the thymus reaches the level of the fourth costal cartilage, whereas superiorly, extensions into the neck are common reflecting the embryonic origin of the thymus.

Ectopic thymus in both humans and mice reflects a failed migration of thymic tissue from third pharyngeal pouch endoderm during organogenesis. Ectopic thymus is usually located anteriorly and deep to the middle third of the sternocleidomastoid muscle and adheres posteriorly to the carotid sheath and extends into the retropharyngeal space (Ahsan et al., 2010). While thymus gland is forming, the parathyroid glands arise simultaneously from the third and fourth pharyngeal pouches and start their migration to a position posterior to the thyroid primordium.

As the thymus proliferates and descends, the local cardiac neural crest mesenchyme controls the pattern and development of the gland. Defective development of cardiac neural crest also results in thymic deficiency as seen in the Di George syndrome.

At the beginning of development, the thymus is a solid epithelial strand composed of densely packed epithelial cells, surrounded by a basal lamina and a vascularized mesenchyme. Later, cortical epithelial cells begin to separate while cells of the medulla remain densely packed. Vascularized mesenchyme transforms into connective septa that invade epithelial strands up to the medulla, subdividing the cortical zone into lobules, but not completely subdividing the medullary zone. During further development, thymic lobules with their well-delimited cortex and medulla become packed tightly together.

The thymus receives its blood supply from branches of the internal thoracic arteries and inferior thyroid arteries. Venous blood from the thymus drains into the brachiocephalic and internal thoracic veins, which communicates above with the inferior thyroid veins. Arterial branches pass directly into the depths of the interlobar septa before entering the thymus at the junction of the cortex and medulla and form capillaries, which distribute into the cortex. Most cortical capillaries loop around at different depths in the cortex and join venous vessels at the cortico-medullary junction. Some continue through the cortex to join larger veins running in the capsule and so leave the thymus.

The cortex is exclusively supplied by capillaries, and the cortico-medullary boundary and the medulla also contain arterioles and venules. There is very little movement of macromolecules from blood to thymic parenchyma across the capillary walls in the cortex, due to the presence of the blood–thymus barrier. This barrier is formed by the continuous blood capillaries in the thymic cortex, capillary basal lamina, basal lamina of epithelial cells, and the epithelial cells. The blood–thymus barrier separates cortical T cells from the blood of cortical vessels, protecting T cells against foreign antigens, completely isolates the thymus cortex, creating a specific microenvironment in which T cells develop into mature T cells (Kato and Schoefl, 1989).

The large medullary vessels are highly permeable to substances in the plasma, and great number of lymphocytes enter the bloodstream by traversing the walls of the postcapillary venules of the cortico-medullary junction and those of the medulla (Lind et al., 2001). Postcapillary venules represent the main route for discharge of mature T cells. Only a small proportion of T cells is carried out from the thymus by the efferent lymphatic vessels. In contrast to the cuboidal endothelium of postcapillary venules of the appendix, Peyer's patches, tonsils, and lymph nodes, the endothelium of thymic postcapillary venules is flattened.

The thymus is an encapsulated gland that undergoes remarkable age-related changes. In relation to body weight, the thymus is largest during embryonic life and in childhood up to the period of puberty. After this, it begins to involute, a process which proceeds gradually and continuously throughout life under normal condition. At birth, the thymus weighs 12 to 15 g. This increases to about 30 to 40 g, at puberty (Hasselblach et al., 1997) after which it begins to decrease in weight, so that at 60 years it weighs only 10 to 15 g. The rate of thymic growth in the child and involution in the adult is extremely variable, and so it is difficult to determine weight appropriate for age (Levine and Rosal, 1978).

Mast cells may be present in large numbers in aged thymuses, where they are largely confined to the inner medulla, septae, and capsule. Although there is a considerable age involution, the thymus remains a functional organ. The thymic involution begins within the cortex, which may disappear completely with aging, whereas medullary persists throughout life. In adults, the thymus is transformed into a mass of adipose tissue (*corpus adiposum thymi*), containing scattered islands of parenchyma consisting mainly of enlarged reticular cells. An accidental involution occurs in response to a wide variety of stimulus, including disease, severe stress, dietary deficiencies, ionizing radiation, injection of colloidal substances, bacterial endotoxin, adrenocorticotrophic hormone, and adrenal and gonadal steroids. Injections of glucocorticoids eliminate as much as 75% of thymocytes within 2 to 3 days. The changes affect both the cortex and the medulla but are most pronounced in the cortex. Under these conditions, the thymus rapidly diminishes in size, due to massive death of cortical small lymphocytes and their destruction by macrophages. The thymus is replenished with fatty areolae that replace its normal lymphoid tissue. Disappearance of thymic structures, however, is not complete, and some islands of functionally competent tissue are still recognizable in senility.

The gland displays a lobuled pattern, with distinct cortical and medullary compartments that is strictly related to its function, namely the production of fully competent circulating T cells bearing the form of the T cell receptor. In hematoxylin–eosin-stained sections, the cortex appears dark blue to purple because of the predominance of lymphocytes (80%–85%), whereas the medulla appears clear (eosinophilic) because of the predominance of the epithelial cells. The cortex contains dense compacted thymocytes that appear as lymphocytes of slightly variable size with scattered, rare mitoses. The lighter-staining medulla contains loosely arranged mature thymocytes and the thymic or Hassall's corpuscles. Formation of Hassall's corpuscles begins with degeneration of an epithelial cell, swelling of its nucleus, cytoplasm, and mitochondria. This cell becomes surrounded by one or more other epithelial cells which are organized circumferentially and connected closely to one another by numerous desmosomes. Keratohyalin granules and numerous tonofilaments appear in central cells. As the innermost cells gradually become distant from blood capillaries, they swell, degenerate, and transform into keratinized and/or necrotic material which often calcifies. Hassall's corpuscles frequently measure 100 μm in diameter, increasing in number and size with age.

Two main cell populations are recognizable (Anderson and Jenkinson, 2001). The stromal population consists of fixed ectodermal-derived, keratin-positive epithelial cells, which form a three-dimensional network occupying the cortex and the medulla. These cells are comprehensively referred to as thymic epithelial reticular cells. Ultrastructural studies of these cells reveal evidence of their epithelial nature such as desmosomes, cytoplasmic tonofilament, and many other organelles found in epithelial cells. Six types of epithelial cells can be identified. Some very large thymic epithelial cells (types 2 and 3) embrace numerous thymocytes and create microenvironment niches in the outer cortex, called thymic nurse cell complexes (Brelinska and Warchol, 1997). Thymic nurse cells express high levels of major histocompatibility complex (MHC) class I and II molecules and contain acidic organelles necessary for antigen processing and presentation. In the medulla, thymic epithelial cells are less interconnected but more compact and heterogeneous.

Thymic epithelial cells secrete cytokines [interleukin (IL)-1, IL-2, IL-3, IL-6, and granulocyte-macrophage colony stimulating factor (GM-CSF)], chemokines (Savino et al., 2002), and thymic hormones and neuropeptides (Mentlein and Kendall, 2000). Moreover, the thymic epithelium is the main source of laminin, fibronectin, and type IV collagen, which are thought to provide preformed routs for thymic migration inside the organ (Savino et al., 2002). Between each epithelial cell and thymocyte, several adhesion points occur, indicating that cell-to-cell contact may be significant in T cell maturation, suggesting a nursing function for epithelial cells (Reike et al., 1995). The large number of highly variable epithelial-like cells and smaller number of thymocytes in the medulla result in less basophilic staining.

Keratin-negative cells include fibroblasts, nonfibroblastic mesenchymal cells, and endothelial cells. Fibroblasts are found in the capsule, perivascular space, and in the medulla but are infrequent in the cortex, except in the involuted thymus. Myoid cells are situated mainly in the medulla and at the cortico-medullary junction. It has been suggested that their contraction might aid the movement of lymphoid cells across or out of the thymus.

The second population constitutes the parenchyma and is composed of thymocytes plus a variety of antigen presenting cells, including interdigitating dendritic cells, macrophages particularly at the corticomedullary junction, and small amounts of B cells. Dendritic cells are involved in shaping and maturating T cells by deleting self-reactive thymocytes to established central tolerance (Varas et al., 2003). Electron microscopic detection of Barbera granules in the cytoplasm of dendritic cells indicates that they express a Langerhans-cells-like phenotype during human ontogeny (Valledeau et al., 2000). Thymic cortical dendritic macrophages have been described (Wakimoto et al., 2008), containing apoptotic thymocytes, express CD8 and MHC II molecules, as well as some dendritic-cell-associated molecules, including fascin, an actin binding protein. They have well-developed processes spreading in the adjacent areas surrounding T cells, are placed in all thymus regions, and are positive for two antibodies, anti-F4/80 and anti-Mac-2 (Liu et al., 2013). The proportion of human thymic dendritic cells remains constant between postnatal, adult, and old humans (Varas et al., 2003). In the cortical region are localized cortical macrophages with flat shape and scanty cytoplasm. They are stained by anti-Mac-2 but not by anti-F4/80 antibodies (Liu et al., 2013).

Thymocyte precursors reach the thymus from the blood. In the embryo, they first come from the rudimentary liver and then from the bone marrow. On entering the gland, they undergo proliferation, lineage commitment and selection, which is largely under the control of thymic epithelial cells.

Two selective processes accompany thymocyte migration, proliferation and differentiation (Sprent and Kishimoto, 2002). The final result is the apoptosis of about 96% of thymocytes, and only 3%–5% becomes fully competent T cells, i.e., cells able to recognize foreign antigens, but unresponsive toward self-antigens that eventually enter the circulation as naïve T cells. Both cortex and medulla compartments are thought to provide selective signals leading to cell survival or death.

A major step forward was made as various cell surface molecules became identifiable with the use of specific antisera. The existence of the Thy-1 antigen on T cells (Reif and Allen, 1964) and the high density of surface Ig on B cells (Raff, 1971) allowed to distinguish and separate T from B cells. Maturation of T cells is accompanied by the sequential acquisition of the various T cell markers. Terminal deoxynucleotidyl transferase is found in prothymocytes and immature thymocytes but is absent in mature T cells (Hale, 2004).

T cells could be further subdivided in T cells which help B cells produce antibody bearing CD4, whereas those which exert cytotoxic functions usually have CD8 molecules. Early thymocytes account for less than 3% of the cells in a mouse thymus and are referred to a double-negative cells, because they bear neither the CD4 nor the CD8 accessory molecules (coreceptors) which characterize mature T cells. These double-negative cells proliferate, rearrange, and express T-cell receptor (TCR) genes and produce in a few days the predominant thymus cell types located in the cortex: the double-positive cell. More than 50% of these cells bear a low concentration of surface TCR. These cells become the mature single positive thymus lymphocytes (CD4-positive or CD8-positive) that localize in the medulla and are indistinguishable from peripheral T cells and express high levels of TCR. The high proliferative rate of thymocytes is paralleled by a massive rate of cell death which affects the vast majority of the double-positive cells (McPhee et al., 1979). E-cadherin is strongly expressed on epithelial cells as well as on the double-negative (CD4-negative or CD8-negative) thymocytes in mice, suggesting its participation in the interactions between these two cell types. However, in human thymus, E-cadherin is expressed only on epithelial cells (Kutlesa et al., 2002).

Thymus-derived cells circulate in blood and lymph for as long as many months in rodents and years in man (Miller and Osoba, 1967). They populated discrete areas of the lymph nodes, spleen, and gut-associated lymphoid structures. Other regions of the lymphoid tissues are not colonized by thymus-derived cells, nor are they depleted of lymphocytes following thymectomy. The notion of thymus-dependent and thymus-independent areas in lymphoid tissues is thus established (Parrott et al., 1966). The thymus-dependent areas in the lymph nodes are the paracortical areas, whereas the thymus-independent areas are the follicles and the medullary cords.

Subcapsular thymocytes express both helper/inducer and suppressor/cytotoxic phenotypes (CD4-positive–CD8-positive or "double-positive" thymocytes). These cells then return to the medulla and, during their passage through specific cortical zones, undergo either positive or negative selection under the guidance of both contact and paracrine signals from the epithelium (Petrie, 2002).

Epithelial cells present thymocytes with an enormous repertoire of self-peptides conjugated to MHC moieties. Positive selection is obtained when these complexes on the surface of thymic cells are recognized by the T cell receptor located on the thymocyte surface. This interaction generates survival signals that rescue thymocytes from apoptosis. By contrast, negative selection occurs when such interactions are too strong or too weak, and apoptosis is promoted by the absence of survival factors. This is called "apoptosis by neglect" (Klein and Kyewski, 2000). Only about 4% of double-positive cells are positively selected to generate mature CD4-positive or CD8-positive cells. Once this phenotype is acquired, thymocytes enter the medulla where they remain for a few days before being released into the peripheral lymphoid pool.

The structure of the medulla is substantially different from that of the cortex. Its cell population is very composite, and the stromal epithelium itself is more compact and less arborized. A close relationship has recently been detected between the architecture of the medullary stroma and the emergence of autoimmune disorders in the mouse. The correct expression of the product of the autoimmune regulator (AIRE) gene correlates with a normal organization of the medullary stroma (Zuklys et al., 2000). AIRE encodes a transcriptional regulator that promotes ectopic expression of a repertoire of transcripts encoding proteins that ordinarily are restricted to differentiated organs residing in the periphery (Mathis and Benoist, 2009). By contrast, mutations in the AIRE gene are responsible for an autoimmune syndrome called APECED (autoimmune polyendocrinopathy-candidiasis-ectodermal dystrophy), characterized by loss of self-tolerance to multiple organs and abnormal structure of the thymic medulla (Ramsey et al., 2002).

Everett et al. (1964) demonstrated that cells deriving from the thymus are both short-lived and long-lived. Long-lived cells in man have a life span upward of 5 years and perhaps over 10 years. In the mouse, their life span is more than 80 days.

2.4 STUDIES OF THE THYMUS IN THE CHICK

In birds, the ventro-lateral part of the pharyngeal endoderm of the third and fourth branchial pouches can be separated by trypsinization from the underlying mesenchyme between the 16- and 30-somite stage. If endoderm is associated with an appropriate mesenchyme, thymic histogenesis proceeds, demonstrating that no intervention of ectoderm is required for thymus differentiation (Le Douarin, 1967). In the quail-chick chimeras, the third and fourth endodermal pouches and subsequently the thymic epithelial cells were surrounded by mesenchymal cells derived from the grafted neural crest (Le Douarin, 1973a,b).

Moore and Owen (1967b) used a sex-chromosome marker system in paired chick embryos joined by vascular anastomoses of chorioallantoic or yolk-sac blood vessels. They demonstrated that chromosome analysis following yolk sac anastomosis at 4–5 days of incubation revealed high levels of chimerism in the thymus. When the anastomosis was established later in the development, only low levels of chimerism were found, suggesting that an inflow of blood-borne stem cells is responsible for lymphoid differentiation in the chick thymus. These results led the authors to formulate the "hematogenous theory of blood forming organ histogenesis." According to Moore and Owen, the source of the blood-borne stem cells that invade the primary lymphoid organs is located in the yolk sac, and it would be attractive to consider the hypothesis of a single cell precursor of all blood cells of both erythroid and lymphocytic series.

When a quail neural tube had been grafted into a chick, the thymic lymphocytes were always of chick type even though the thymus mesenchyme was made up of quail cells, excluding the mesenchymal origin of lymphocyte precursor cells. Moreover, the potentiality of the endoderm to give rise to lymphocytes was tested by transplanting the third and fourth pharyngeal pouches endoderm of a 15- and 30-somite quail embryo into the somatopleure of a chick (Le Douarin and Jotereau, 1973).

As total body irradiation destroys lymphoid tissues, Miller predicted that recovery of immune functions following irradiation would be thymus-dependent. Adult mice were thymectomized and subjected to total body irradiation. To prevent death, they were given bone marrow cells, a source of hematopoietic stem cells. Control, nonthymectomized mice treated in this way recovered normal lymphoid functions within 6 to 8 weeks. The thymectomized mice did not (Miller, 1962a, 1963). In the adult, therefore, the thymus is still required to reestablish defense mechanisms depleted as a result of some accident or disease.

Immune functions can be restored to animals thymectomized at birth or thymectomized and irradiated in adult life by infusing lymphocytes or implanting thymus tissue. Miller investigated the effect of injecting lymphoid cells into neonatally thymectomized mice and found that

1. Syngeneic thymus cells from 1-day-old mice given intravenously to newborn mice immediately after thymectomy did not prevent runting, lymphoid atrophy, or immunological failure (Miller, 1962b).
2. Syngeneic lymphoid cells from 8-week-old mice presensitized against actinic keratosis (AK) skin, on injection into 10-week-old neonatal thymectomized C3H mice carrying healthy Ak skin grafts for over 1 month conferred adoptive immunity. The Ak skin was rejected within 12 days, and the mice showed evidence of immunity to a second-set graft (Miller et al., 1964).

3. Allogeneic lymphoid cells from 2-month-old mice caused a severe graft-versus-host (GVH) reaction when injected intravenously into newborn mice immediately after thymectomy (Miller, 1962a).

Lymphocytes restored immune capabilities, but only if the donor was syngeneic (Miller, 1961). If it was of a different strain, the thymectomized host wasted and died, because injected lymphocytes, being immunologically competent, reacted against the foreign tissues of their host and brought about a fatal GVH reaction.

Neonatal thymectomized mice, implanted with syngeneic thymus tissue soon after birth, developed a normal immune system. When grafted with foreign thymus tissue, they were specifically tolerant of thymus-donor type skin only (Miller, 1962b, 1963).

This finding led Miller to postulate, "when one is inducing a state of immunological tolerance in a newly born animal, one is in effect performing a selective or immunological thymectomy" (Miller, 1962b). In other words, the precursors of thymic lymphocytes differentiating in the presence of foreign cells and with specificities for the foreign antigens would be deleted, implying that the thymus might be the site where self–nonself-discrimination occurs and self-tolerance is imposed.

2.5 IDIOPATHIC ACQUIRED AGAMMAGLOBULINEMIA ASSOCIATED WITH THYMOMA

In 1953, his colleague Richard Varco asked Good to consult on a 54-year-old male patient who had initially presented to his chest clinic in June 1951. The patient complained of having suffered at least 17 bouts of pneumonia during the previous 8 years and a pronounced susceptibility to infection, which had increased, concomitant with the appearance and extirpation of a benign thymoma, occupying almost the entire thymic gland (Mac Lean et al., 1956).

The interesting thing to Good about this patient was that he also carried a diagnosis of "acquired agammaglobulinemia," a markedly deficient ability to produce antibodies and significant deficits of all or most of the cell-mediated immunities. Surgical removal of the tumor, which was primarily an epithelial stromal overgrowth of the thymus, did not correct the immunodeficiencies in this patient. Since that, seven cases of the combined occurrence of these two disorders were reported

(Ramos, 1956; Martin et al., 1956; Lambie et al., 1957; Gafni et al., 1960), and in no instance, removal of the thymic tumor restored immunologic function or the protein deficit.

Good and Varco (1955) described a new syndrome that would carry his name: "Good's syndrome: thymoma with immunodeficiency." The clinical characteristics of Good's syndrome were increased susceptibility to bacterial infections with encapsulated organisms and opportunistic viral and fungal infections. Subsequently, Good saw several patients with thymic tumors, which regularly presented with immunodeficiencies, leukopenia, lymphopenia, and eosinophylopenia. Plasma cells, however, were not completely absent: the patient was severely hypogammaglobulinemic rather than agammaglobulinemic.

The association of thymoma with the profound and broadly based immunodeficiency provoked Good's group the question what role the thymus plays in immunity.

Good (Good, 1954; Bridges and Good, 1960) and others (Janeway et al., 1953; Gitlin et al., 1956) found that the patients lacked all of the subsequently described immunoglobulins (Bridges and Good, 1960). These patients were found not to have plasma cells or germinal centers in their hematopoietic and lymphoid tissues, whereas they possessed circulating lymphocytes in normal numbers (Good, 1955).

Good decided to investigate the possibility that the thymus had something to do with adaptive immunity, and under his direction, Zak and Mac Lean performed thymectomies on 4–5-week-old rabbits, but they found that thymectomy had no demonstrable effects on the antibody response (Mac Lean et al., 1956, 1957). In the discussion of the second paper, the authors noted that, although their laboratory investigation had not led to discover the exact function of thymus, they believed that their patient represented an experiment of nature that suggested that the thymus does, indeed, play a crucial role in immunity.

2.6 THE EFFECTS OF NEONATALLY THYMECTOMY

In the mouse and other rodents, immunologic depression is profound after thymectomy in neonatal animals, resulting in considerable depression of antibody production, plus deficient transplantation immunity and delayed-type hypersensitivity (Waksman et al., 1962). Speculation on the reason for immunologic failure following neonatal thymectomy has centered in the thymus as a source of cells or humoral factors essential to normal lymphoid development and immunologic maturation.

At the University of Wisconsin, a second group of investigators was engaged in endocrinologic studies which led to the first experiments on neonatally thymectomy in rabbits. Three independent groups of experiments showed that neonatal thymectomy has a significant effect on immunologic reactivity: (1) The studies of Fichtelius et al. (1961) in young guinea pigs showed that the depression of antibody response is slight, but significant; (2) The experiments of Archer, Good and coworkers in rabbits (Archer and Pierce, 1961; Archer et al., 1962; Good et al., 1962) and mice (Good et al., 1962; Martinez et al., 1962a,b; Dalmasso et al., 1963); and (3) The studies by Miller at the Chester Beatty Research Institute in London (Miller, 1961, 1962a,b, 1963; Miller et al., 1964).

In rabbits, the effects of neonatal thymectomy on antibody production were variable both from animal to animal and antigen to antigen (Archer et al., 1962). In the mouse, transplantation immunity was sufficient affected by neonatal thymectomy to permit skin transplants across the H-2 histocompatibility barrier and even across species barriers in some instances, and antibody production to certain antigens was almost entirely eliminated (Martinez et al., 1962a,b; Miller, 1962a, b; Papermaster et al., 1962c).

Good wrote, "the simultaneous occurrence of acquired agammaglobulinemia and benign thymoma in a human being, suggested that the thymus might participate in the control of antibody formation. [...] It still seems likely that some essential relationship exists between the thymic tumor and the acquisition of an acquired agammaglobulinemia. A second case of acquired agammaglobulinemia with thymoma presents itself and strengthens the conviction that the two phenomena are related in some essential manner."

Parrott and East (1964) showed that the effect on antibody production was a quantitative one, and that with such potent antigens as hemocyanin and pneumococcal polysaccharides, many of the neonatally thymectomized mice produced significant amounts of antibody.

Neonatally thymectomized mice were particularly vulnerable to homologous disease when injected with parent strain lymphoid cells

(Parrott, 1962). Their lymphoid tissues showed minimal lymphoid development, and their circulating lymphocyte levels were greatly reduced, although most strains have adequate levels of gammaglobulins and appreciable, if not entirely normal, number of plasma cells. The cells of spleen and lymph nodes of neonatally thymectomized mice were very low in immunologic activity and splenomegaly, and other manifestations of graft versus host (GVH) reactivity were not evident unless the dosage of cells was multiplied several times (Dalmasso et al., 1963).

Rabbits thymectomized during the neonatal period revealed gross deficiencies in the distribution of T lymphocytes. Parrott et al. (1966) showed that thymic derived lymphocytes occupied the paracortical regions in the lymph nodes and periarterial regions in the spleen, the so-called thymus-dependent regions (Cooper et al., 1966). Other specialized areas in the lymph nodes are located in the far cortical regions, the B-zones, where germinal center developed.

Overall, these experimental data indicate that the mouse has one primary central lymphoid organ, the thymus, and that this is a key source of cells or humoral substances, or both, necessary to normal maturation of the peripheral lymphoid tissues and to normal development of immunologic capabilities. The fact that neonatally thymectomized mouse retained a small measure of immunologic reactivity suggests that the thymus' influence on immunologic development may already have been excised to some extent before birth.

2.7 THE RESTORATIVE EFFECT OF THYMOMAS OR THYMUS GRAFTS ON THYMIC FUNCTION

Stutman in Good's laboratory demonstrated that nonlymphoid thymomas induced restoration of immunological functions in neonatally thymectomized mice (Stutman et al., 1967), and that when thymomas were grafted into allogenic hosts, immunological restoration was mediated by lymphoid cells of host type (Stutman et al., 1968). Comparable results were obtained with free thymus grafts (Stutman et al., 1969a, 1969b). When the treatment of neonatally thymectomized host was delayed after neonatal thymectomy, a decrease in the restorative capacity was observed (Stutman et al., 1969a, 1969b). These results indicate that a population of cells in thymectomized hosts capable of responding to the action of thymus or thymomas decreased progressively with time after neonatal thymectomy (Stutman et al., 1969a, 1969b).

2.8 REMOVAL OF EITHER THE THYMUS OR BURSA OF FABRICIUS

The bursa of Fabricius and the thymus are "central lymphoid organs" in the chicken, essential to the ontogenetic development of adaptive immunity in that species. Studies by Papermaster et al. (1962a,b) in Good's laboratory indicated that bursectomy in the newly hatched chicks did not completely abolish immunologic potential in the adult animal but rather produces a striking quantitative reduction insufficient to eliminate the homograft reaction. The failure of thymectomy in newly hatched chicks to alter the immunologic potential of the maturing animal probably only reflected the participation of the bursa of Fabricius in the development of full immunologic capacity.

In 1963, Max Cooper started a long series of experiments in Good's laboratory. He removed either the thymus or bursa of Fabricius from some newly hatched chicks, and both from others. Then, to destroy all peripheral lymphoid components, he subjected the chicks to intense x-irradiation just below the lethal level, to destroy cells that might have seeded earlier from the thymus and bursa or that could have been influenced by postulated thymic and bursal humoral factors. Finally, Cooper waited several weeks until the experimental and irradiated control animals recovered from the irradiation effects.

Results showed that the bursectomized and irradiated birds were completely devoid of germinal centers, plasma cells and the capacity to make antibodies, yet they had perfectly normal development of thymocytes and lymphocytes elsewhere in the body that mediated cellular immune reactions (Cooper et al., 1965, 1966). On the other hand, thymectomized and irradiated animals were deficient in lymphocytes that mediated cellular immunity as assessed by skin graft rejection, delayed-type hypersensitivity, and graft versus host assays, but they still produced germinal centers, plasma cells, and circulating immunoglobulins (Cooper et al., 1965, 1966). Like thymectomized mice, the thymectomized and irradiated birds had impaired antibody responsiveness to antigens. Birds subjected to combined thymectomy, bursectomy, and irradiation had severe cellular and humoral immune system deficiency (Cooper et al., 1965, 1966).

Cooper et al. (1965) postulated that a lymphoid stem cell population exists that is induced to differentiate along two distinct and separate cell lines related to two central lymphoid organs. In birds, this developmental influence is exercised by the thymus and the bursa of Fabricius. Removal of one or both in the early posthatching period has strikingly different influences on immunologic function in the maturing animals (Cooper et al., 1965, 1966). The thymus in the chicken functions exactly as does the thymus of the mouse. It represents the site of differentiation of a population of lymphocytes that subserve largely the functions of cell-mediated immunity.

Overall, these data indicate that at some point differentiation along two distinctly different pathways occurs within the lymphoid system and that the critical point seems to focus about two separate central lymphoid organs.

2.9 THE IMPORTANCE OF A FUNCTIONAL THYMUS IN HUMAN

There are several clinical observations which serve to underline the importance of a functional thymus to immunologic development in man.

Good said that one morning in 1952, he "opened his green pediatric journal" and found an article by Odgen Bruton describing the electrophoretic pattern of an 8-year-old patient, seemingly, devoid of gammaglobulin, i.e., agammaglobulanemic. Bruton's initial reports were published in 1952 (Bruton, 1952; Bruton et al., 1952). He described a patient for whom life was one severe life-threatening bacterial infection after another. Immediately after Bruton's description of the association of immunodeficiency with agammaglobulinemia, Janeway's group in Boston (Janeway et al., 1953; Gitlin et al., 1956) and Good's group in Minneapolis (Good and Varco, 1955; Good and Zak, 1956) launched studies with on series of patients with X-linked agammaglobulinemia (XLA). XLA was one of the first recognized primary immunodeficiencies of man. Most of these patients exhibited normal delayed-type hypersensitivity and rejected primary as well as secondary skin allografts. In these patients, thymus was found almost normal, and thymus-dependent lymphocytes were also normal as well as their functions. Life of these agammaglobulinemic children involved a succession of life-threatening episodes of infection usually caused by *Streptococcus pneumonii*, *Hemophilus influenzae*, meningococci, or other high-grade encapsulated pyogenic pathogens including *Pseudomonas aeruginosa*. Furthermore, these patients who did not produce antibodies did not have normal adaptive immune responses that involved a normal ability to express, terminate, and resist recurrence childhood exanthems, e.g., chicken pox and vaccinia. They also resisted bacille Calmette-Guérin (BCG) infection quite normally (Good and Varco, 1955; Good, 1954, 1955; Good and Zak, 1956).

However, a number of these patients had vestigial thymus tissue weighing less than a gram in most instances, and almost completely lacking in lymphoid cells (Good et al., 1964). In several of these patients, the thymus was not only very small and very deficient in lymphoid development but showed a failure of descensus.

The Di George third and fourth pharyngeal pouch syndrome patients had absence or deficiency of all cell-mediated immunological functions. They were also somewhat deficient in antibody production, but not as deficient as the patients with Bruton's XLA. They had low set, crumpled, abnormally rotated ears and characteristic face with a small mandible, short philtrum, and a small low-shaped mouth (Conley et al., 1979). Di George patients had severe deficiencies of small T lymphocytes and profound deficiencies of all cell-mediated immunities, including delayed allergies, deficient allograft immunities and deficiencies in resistance to viruses, fungi, and opportunistic infections. Having defined the Di George syndrome as selective deficiency of T cell development due to failure of differentiation of thymus, it seemed likely that the abnormality should be correctable by thymus transplants (Cleveland et al., 1968; August et al., 1970; Biggar et al., 1972). The athymic children described by Di George, who lacked lymphoid cells in the deep cortical areas of the nodes but not at the peripheral areas, seemed the equivalent of the neonatally thymectomized mice and chickens.

REFERENCES

Ahsan, F., Allison, R., White, J., 2010. Ectopic cervical thymus: case report and review of pathogenesis and management. J. Laryngol. Otol. 124, 694–697.

Anderson, G., Jenkinson, E.J., 2001. Lymphostromal interactions in thymic development and function. Nat. Rev. Immunol. 1, 31–40.

Archer, O.K., Pierce, J.C., 1961. Role of thymus in the development of the immune response. Fed. Proc. 20, 26.

Archer, O.K., Pierce, J.C., Papermaster, B.W., et al., 1962. Reduced antibody response in thymectomized rabbits. Nature 191, 192–193.

Auerbach, R., 1961. Experimental analysis of the origin of cell types in the development of the mouse thymus. Dev. Biol. 3, 336–354.

Auerbach, R., 1963. Developmental studies of mouse thymus and spleen. Natl. Cancer. Inst. Monogr. 11, 23–33.

August, C.S., Rosen, F.S., Miller, R.M., et al., 1970. Implantation of a fetal thymus, restoring immunological competence in a patient with thymus aplasia (Di George syndrome). Lancet. ii, 1210–1211.

Biggar, W.D., Stutman, O., Good, R.A., 1972. Morphological and functional studies of fetal thymus transplants in mice. J. Exp. Med. 135, 793–807.

Billingham, R.E., Brent, L., Medawar, P.B., 1956. Quantitative studies on tissue transplantation immunity. III. Actively acquired tolerance. Philos. Trans. R. Soc. London 239 B, 357–412.

Brelinska, R., Warchol, J.B., 1997. Thymic nurse cells: their functional ultrastructure. Microsci. Res. Techn. 38, 250–266.

Bridges, R.A., Good, R.A., 1960. Connective tissue diseases and certain serum protein components in patients with agammaglobulinemia. Ann. N. Y. Acad. Sci. 86, 1089–1097.

Bruton, O.C., 1952. Agammaglobulinemia. Pediatrics. 9, 722–728.

Bruton, O.C., Apt, L., Gitlin, D., et al., 1952. Absence of serum gammaglobulins. Am. J. Dis. Child. 84, 632–636.

Cleveland, W.W., Fogel, B.S., Brown, W.T., et al., 1968. Foetal thymus transplant in a case of Di George syndrome. Lancet. ii, 1211–1214.

Conley, M.E., Beckwith, J.B., Mancer, J.F.K., et al., 1979. The spectrum of the Di George syndrome. J. Pediatr. 94, 883–890.

Cooper, M.D., Peterson, R.D.A., Good, R.A., 1965. Delineation of the thymic and bursal lymphoid systems in the chicken. Nature. 205, 143–146.

Cooper, M.D., Peterson, R.D.A., South, M.A., et al., 1966. The functions of the thymus system and the bursa system in the chicken. J. Exp. Med. 123, 75–102.

Crivellato, E., Vacca, A., Ribatti, D., 2004. Setting the stage: an anatomists's view of the immune system. Trends. Immunol. 25, 201–207.

Dalmasso, A.P., Martinez, C., Sjodin, K., et al., 1963. Studies on the role of the thymus in immunobiology. Reconstruction of immunologic capacity in mice thymectomized at birth. J. Exp. Med. 118, 1089–1109.

Everett, N.B., Caffrey, R.W., Rieke, W.O., 1964. Recirculation of lymphocytes. Ann. N. Y. Acad. Sci. 113, 887–897.

Feldman, M., Globerson, A., 1964. The role of thymus in restoring immunological reactivity and lymphoid cell differentiation in X-irradiated adult mice. Ann. N. Y. Acad. Sci. 120, 182–190.

Fichtelius, K.E., Laurell, G., Philipsson, L., 1961. The influence of thymectomy on antibody formation. Acta Path Microbiol. Scand. 51, 81–86.

Gafni, J., Michaeli, D., Heller, H., 1960. Idiopathic acquired agammaglobulinemia associated with thymoma; report of two cases and review of the literature. New Engl. J. Med. 263, 536–541.

Gitlin, D., Hitzig, W.H., Janeway, C.A., 1956. Multiple serum protein deficiencies in congenital and acquired agammaglobullinemia. J. Clin. Invest. 35, 1199–1204.

Good, R.A., 1954. Agammaglobulinemia: a provocative experiment of nature. Bull. Univ. Minn. Hosp. Minn. Med. Fdn 26, 1–19.

Good, R.A., 1955. Studies on agammaglobulinemia. II. Failure of plasma cell formation in the bone marrow and lymph nodes of patients with agammaglobulinemia. J. Lab. Clin. Med. 46, 167–181.

Good, R.A., Campbell, B., 1945. Potentiating effect of anaphylactic and histamine shock upon Herpes simplex virus infection in rabbits. Proc. Soc. Exp. Biol. Med. 59, 305–306.

Good, R.A., Varco, R.L., 1955. A clinical and experimental study of agammaglobulinemia. Lancet. 75, 245–271.

Good, R.A., Zak, S.J., 1956. Disturbances in gamma globulin synthesis as experiments of nature. Pediatric 18, 109–149.

Good, R.A., Dalmasso, A.P., Martinez, C., et al., 1962. The role of the thymus in development of immunologic capacity in rabbits and mice. J. Exp. Med. 116, 773–796.

Good, R.A., Martinez, C., Gabrielsen, A.E., 1964. Clinical considerations of the thymus in immunobiology. In: Good, R.A., Gabrielsen, A.E. (Eds.), The Thymus in Immunobiology. Hoeber-Harper, New York, pp. 3–47.

Gowans, J.L., Mc Gregor, D.D., Cowen, D.M., 1962. Initiation of immune response by small lymphocytes. Nature. 196, 651–655.

Hale, L.P., 2004. Histologic and molecular assessment of human thymus. Ann. Diagn. Pathol. 8, 50–60.

Hasselblach, H., Jeppesen, D.L., Ersbroll, A.K., 1997. Thymus size evaluated by sonography. A longitudinal study on infants during the first year of life. Acta. Radiol. 38, 222–227.

Janeway, C.A., Apt, L., Gitlin, D., 1953. Agammaglobulinemia. Trans. Assoc. Am. Physicians. 66, 200–202.

Kato, S., Schoefl, G.I., 1989. Microvascolature of normal and involute mouse thymus. Light and electron microscopic study. Acta Anat. 135, 1–11.

Klein, L., Kyewski, B., 2000. Self-antigen presentation by thymic stromal cells: a subtle division of labor. Curr. Opin. Immunol. 12, 179–186.

Kutlesa, S., Wessels, J.T., Speiser, A., et al., 2002. E-cadherin-mediated interactions of thymic epithelial cells with CD103^{+} thymocytes lead to enhanced thymocyte cell proliferation. J. Cell. Sci. 115, 4505–4515.

Lambie, A.T., Burrows, B.A., Sommers, S.C., 1957. Clinicopathologic conference: refractory anemia, agammaglobulinemia, and mediastinic tumor. Am. J. Clin. Pathol. 27, 444–454.

Le Douarin, N., 1967. Early determination of the anlagen of the thyroid and thymus glands in the chick embryo. C. R. Acad. Sci. 264, 940–942.

Le Douarin, N., 1973a. A biological cell labeling technique and its use in experimental embryology. Dev. Biol. 30, 217–222.

Le Douarin, N., 1973b. A Feulgen-positive nucleolus. Exp. Cell. Res. 77, 459–468.

Le Douarin, N.M., Jotereau, F.V., 1973. Origin and renewal of lymphocytes in avian embryo thymuses studied in interspecific combinations. Nat. New Biol. 246, 25–27.

Le Douarin, N.M., Jotereau, F.V., 1975. Tracing of cells of the avian thymus through embryonic life in interspecific chimeras. J. Exp. Med. 142, 17–39.

Levine, G.D., Rosal, J., 1978. Thymic hyperplasia and neoplasia. A review of current concepts. Hum. Pathol. 9, 495–515.

Lind, E., Prockop, S.E., Porritt, H.E., et al., 2001. Mapping precursor movement through postnatal thymus reveals specific microenvironment supporting defined stages of early lymphoid development. J. Exp. Med. 194, 127–134.

Liu, L.A., Lang, Ze, Li, Y., et al., 2013. Composition and characteristics of distinct macrophage subpopulations in the mouse thymus. Mol. Med. Rep. 7, 1850–1854.

Mac Lean, L.D., Zak, S.J., Varco, R.L., et al., 1956. Thymic tumor and acquired agammaglobulinemia: a clinical and experimental study of the immune response. Surgery 40, 1010–1017.

Mac Lean, L.D., Zak, S.J., et al., 1957. The role of the thymus in antibody production: an experimental study of the immune response in thymectomized rabbits. Transp. Bull. 4, 21–22.

Martin, C.M., Gordon, R.S., McCullough, N.B., 1956. Acquired agammaglobulinemia in an adult; report of a case, with clinical and experimental studies. New Engl. J. Med. 254, 449–462.

Martinez, C., Kersey, J., Papermaster, B.W., et al., 1962a. Skin homografts survival in thymectomized mice. Proc. Soc. Exp. Biol. Med. 109, 193–196.

Martinez, C., Dalmasso, A., Good, R.A., 1962b. Acceptance of tumour homografts by thymectomized mice. Nature. 194, 1289–1290.

Mathis, D., Benoist, C., 2009. AIRE. Annu. Rev. Immunol. 27, 287–312.

McPhee, D., Pye, J., Shortman, K., 1979. The differentiation of T lymphocytes. V. Evidence for intrathymic death of most thymocytes. Thymus 1, 151–162.

Mentlein, R., Kendall, M.D., 2000. The brain and thymus have much in common: a functional analysis of their microenvironments. Immunol. Today 21, 133–140.

Miller, J.F.A.P., 1961. Immunological function of the thymus. Lancet ii, 748–749.

Miller, J.F.A.P., 1962a. Role of the thymus in transplantation immunity. Ann. N. Y. Acad. Sci. 99, 340–354.

Miller, J.F.A.P., 1962b. Effect of neonatal thymectomy on the immunologic responsiveness of the mouse. Proc. R. Soc. (London), Series B 156, 415–428.

Miller, J.F.A.P., 1963. Role of the thymus in immunity. Br. Med. J. 2, 459–464.

Miller, J.F.A.P., Osoba, B., 1967. Current concepts of the immunological function of the thymus. Physiol. Rev. 47, 437–520.

Miller, J.F.A.P., Leuchars, E., Cross, A.M., et al., 1964. Immunological role of the thymus in radiation chimeras. Ann. N. Y. Acad. Sci. 120, 205–217.

Moore, M.A., Owen, J.J.T., 1967b. Experimental studies on the development of the thymus. J. Exp. Med. 126, 715–723.

Papermaster, B.W., Dalmasso, A.P., Martinez, C., et al., 1962a. Suppression of antibody forming capacity with thymectomy in the mouse. Proc. Soc. Exp. Biol. Med. 111, 41–43.

Papermaster, B.W., Friedman, D.L., Good, R.A., 1962b. Relationship of the bursa of Fabricius to immunologic responsiveness and homograft immunity in the chicken. Proc. Soc. Exp. Biol. Med. 110, 62–65.

Papermaster, B.W., Bradley, S.G., Watson, D.W., et al., 1962c. Antibody-producing capacity of adult chicken spleen cells in newly hatched chicks. A study of sources for variation in a homologous cell transfer system. J. Exp. Med. 115, 1191–1211.

Parrott, D.M.W., 1962. Strain variation in mortality and runt disease in mice thymectomized at birth. Transplant. Bull. 29, 102–104.

Parrott, D.M.V., East, J., 1964. Studies on a fatal wasting syndrome of mice thymectomized at birth. In: Good, R.A., Gabrielsen, A.E. (Eds.), The Thymus in Immunobiology. Hoeber-Harper, New York, pp. 523–540.

Parrott, D.M.W., De Sousa, M.A.B., East, J., 1966. Thymus-dependent areas in the lymphoid organs of neonatally thymectomized mice. J. Exp. Med. 23, 191–204.

Petrie, H.T., 2002. Role of thymic organ structure and stromal composition in steady-state postnatal T-cell production. Immunol. Rev. 189, 8–19.

Raff, M.C., 1971. Surface antigenic markers for distinguish T and B lymphocytes in mice. Transpl. Rev. 6, 52–80.

Ramos, A.J., 1956. Presentation of a case, with discussion by V. Loeb. J. Am. Med. Assoc. 160, 1317–1319.

Ramsey, C., Bukrinsky, A., Peltonen, L., 2002. Systematic mutagenesis of the functional domains of AIRE reveals their role in intracellular targeting. Hum. Mol. Genet. 11, 3299–3308.

Reif, A.E., Allen, J.M., 1964. The AKR thymic antigen and its distribution in leukemias and nervous tissue. J. Exp. Med. 120, 413–433.

Reike, T., Penninger, J., Romani, N., et al., 1995. Chicken thymic nurse cells: an overview. Dev. Comp. Immunol. 19, 281–289.

Savino, W., Mendes-da-Cruz, D.A., Silva, J.S., et al., 2002. Intrathymic T-cell migration: a combinatorial interplay of extracellular matrix and chemokines?. Trends Immunol. 23, 305–313.

Sprent, J., Kishimoto, H., 2002. The thymus and negative selection. Immunol. Rev. 185, 126–135.

Stutman, O., Yunis, E.J., Good, R.A., 1967. Functional activity of a chemically induced thymic sarcoma. Lancet. i, 1120–1123.

Stutman, O., Yunis, E.J., Good, R.A., 1968. Carcinogen-induced tumors of the thymus. I. Restoration of neonatally thymectomized mice with a functional thymoma. J. Natl. Cancer Inst. 41, 1431–1452.

Stutman, O., Yunis, E.J., Good, R.A., 1969a. Tolerance induction with thymus grafts in neonatally thymectomized mice. J. Immunol. 103, 92–99.

Stutman, O., Yunis, E.J., Good, R.A., 1969b. Carcinogen-induced tumors of the thymus. IV. Humoral influences of normal thymus and functional thymomas and influence of postthymectomy period of restoration. J. Exp. Med. 130, 809–819.

Valledeau, J., Ravel, O., DeZutter-Dambuyant, C., et al., 2000. Langerin, a novel C type lectin specific to Langherans cells, is an endocytic receptor that induces the formation of Birbeck granules. Immunity. 12, 71–81.

Varas, A., Sacedon, R., Hernandez–Lopez, C., et al., 2003. Age-dependent changes in thymic macrophages and dendritic cells. Microsc. Res. Technol. 62, 501–507.

Wakimoto, T., Tomisaka, R., Nishikawa, Y., et al., 2008. Identification and characterization of human thymic cortical dendritic macrophages that may act as professional scavangers of apoptotic thymocytes. Immunobiology. 213, 837–847.

Waksman, B.H., Arnason, B.C., Jankovic, B.D., 1962. Role of the thymus in immune reactions in rats. III. Changes in the lymphoid organs of the thymectomized rats. J. Exp. Med. 116, 187–206.

Zuklys, S., Balciunaite, G., Agarwak, A., et al., 2000. Normal thymic architecture and negative selection are associated with Aire expression, the gene defective in autoimmune-polyendocrinopathy-candidiasis-ectodermal dystrophy (APECED). J. Immunol. 165, 1976–1983.

FURTHER READING

Moore, M.A., Owen, J.J.T., 1967a. Stem cell migration in developing myeloid and lymphoid systems. Lancet 2, 658–659.

CHAPTER 3

The Discovery of the Blood–Thymus Barrier

3.1 THE CONCEPT OF BARRIER BETWEEN BLOOD AND TISSUES

The original concept of a barrier preventing the movement of certain materials between the blood and the adult brain stem from studies of dye injections made into circulation. In 1878, Paul Ehrlich obtained his doctorate in medicine by means of a dissertation on the theory and practice of staining animal tissues. In the same year, he was appointed assistant to Professor Friedrich Theodor von Frerichs at the Berlin Medical Clinic, who gave him every facility to continue his work with these dyes and the staining of tissues with them.

In 1885, Ehrlich reported that after parental injection in adult animals of a variety of vital dyes, practically all animal organs were stained, except the brain and spinal cord. An early conclusion from these experiments was that the specific feature of the central nervous system (CNS) was a lack or low affinity for vital dyes. Although Ehrlich himself described the observation that after intravenous application of some aniline dyes, most of the animal tissues were stained with the exception of the CNS, he thought that this difference was due to different binding affinities.

Common examples for barriers are the blood–brain, the blood–placenta, the blood–retina, the blood–testis, and the blood–thymus barrier. The barriers have a defined anatomic substrate: for the blood–brain, the inner blood–retina, and the blood–thymus barrier, it is the endothelium; for the blood–placenta, the outer blood–retina, the blood–testis, and the blood–thymus barrier, these are epithelial cells near the capillary wall. Epithelia with barrier function have dense intercellular junctions and few pinocytotic vesicles and express many transporters for the selective transport and for the exchange of molecules.

Milestones in Immunology. DOI: http://dx.doi.org/10.1016/B978-0-12-811313-4.00003-6

3.2 HORSERADISH PEROXIDASE AS A MARKER OF VASCULAR PERMEABILITY

Horseradish peroxidase (HRP) is a glycoprotein with a molecular weight of 40,000, which can be demonstrated at both light and electron microscopic level by cytochemical reactions. HRP injected intravenously into mice passed freely out of the capillaries in cardiac and skeletal muscle (Karnovsky, 1967). Since the introduction of the HRP as tracer, the number of hemoproteins used as enzymatic probes has been largely augmented towards both lower and higher molecular weight; the spectrum ranging from 1500 (heme-octapeptide) to 240,000 Da (catalase).

One of the most important contributions of Morris Karnovsky was the extension of HRP tracer method to both light and electron microscopic level, by introducing diaminobenzidine (DAB) as an electron donor. HRP oxides DAB in the presence of H_2O_2 and converts it into an insoluble polymer, which is detectable by light and electron microscope. When HRP is injected into the bloodstream, its pathways can be followed with DAB reaction.

As Karnovsky (2006) wrote in his autobiographic paper: "Werner Straus (1957) had used HRP to study endocytic electron uptake at the light microscopic level, but benzidine, the electron donor in the peroxidase reaction used at that time to reveal the site of the enzyme, did not yield a sufficiently electron-opaque reaction product suitable for electron microscopy. (...) I thought that a suitable substrate (electron donor) could be formulated, one that would yield an insoluble reaction product that would reduce osmium tetroxide and bind the reduced osmium. Modified benzidine, with additional amino groups, seemed to be a suitable candidate, DAB was the answer. (...) In the DAB-peroxidase reaction, HRP oxidizes DAB in the presence of H_2O_2 and converts it into an insoluble brown polymer; this polymer causes reduction of added osmium tetroxide, and the reduced osmium forms an insoluble electron-opaque precipitate at the site of the HRP. Thus, when HRP is injected into the bloodstream, the pathways that it takes to reach the tissues can be followed by fixing tissues at various times after injection and by performing the DAB reaction. Furthermore, HRP can be cross-linked to antibodies and other proteins for localization studies in vivo and in vitro."

In the first paper introducing this technique, Karnovsky studied the passage of HRP through the glomerulus in the urine (Graham and

Karnovsky, 1966). Reese and Karnovsky (1967) showed for the first time at ultrastructural level that the endothelium of mouse cerebral capillaries constitutes a structural barrier to HRP. This barrier is composed of the plasma membrane and the cell body of endothelial cells and of tight junctions between adjacent cells. The tight junctions completely obliterate the narrow cleft between adjacent cells forming a continuous belts or rows of zonulae occludentes. Reese and Karnovsky (1967) found that HRP was able to enter the interendothelial spaces only up to, but not beyond, the first luminal interendothelial tight junctions in cerebral capillaries. Unlike muscle capillaries, these tight junctions appeared to be continuous, and pinocytotic vesicles were uncommon and not involved in the transport. Moreover, a relatively scarce number of vesicles in the cerebral adult endothelia might be one of the morphological features of the functioning blood-brain barrier (BBB), whereas the richness in vacuoles, vesicles, and luminal and abluminal expansions and invaginations displayed by endothelial cells of noncerebral microvessels could be the indication of a high endo-exocytotic activity allowing transendothelial transport.

The most prominent feature of the blood–brain barrier is the presence of complex tight junctions between CNS endothelial cells, which establish a high electrical resistance across the endothelial barrier. In 1968, Schneeberger-Keeley and Karnovsky, by means of HRP, defined the alveolar–capillary barrier, as residing at the tight junctions between the alveolar epithelial cells.

3.3 THE BLOOD–THYMUS BARRIER

The existence of a blood–thymus barrier was proposed for the first in time in 1961 by Marshall and White (1961), and demonstrated morphologically by Clark (1963) and Weiss (1963). Weiss studied the thymus of newborn and young adult mice by means of electron microscope. He demonstrated that in the thymus cortex a cellular pathway from the lumen outside the capillaries was established by means of cytoplasmic processes of the endothelium and the adventitial reticular cells. Moreover, after intravenous injection of thorium dioxide, the tracer was found in the vessel wall, and, to a limited degree, in the surrounding tissue. Weiss (1963) concluded, "These vessels resemble both the fine vessels in the central nervous system where a blood–brain barrier is present, and the terminal arterial vessels in the spleen".

The most clear morphological evidence concerning the existence of the blood–thymus barrier may be attributed to the collaborative work of two scientists, Morris Karnovsky and Elio Raviola (Figs. 3.1 and 3.2). At present, Morris Karnovsky is "Shattuck Professor of Pathological Anatomy," Emeritus, and Elio Raviola, "Bullard Professor of Neurobiology," Emeritus, both at the "Harvard Medical School," Boston, USA. Other areas of interest of Karnovsky are the oxidative metabolism of activated leukocytes and osteoclasts and lipid domains in cell membranes, and the reaction of blood vessels to injury and in transplants. The main focus of Raviola's laboratory is the biology of the eye, including research into the role of visual experience in postnatal eye development and studies of how the retina of mammals analyzes the visual world and encodes information about it for sending to the brain.

The blood–thymus barrier is a functional and selective barrier separating T-lymphocytes from blood and cortical capillaries. In the cortex, capillaries, which are rarely fenestrated, form the barrier together with perivascular lymphocytes, macrophages, and reticular epithelial cells. Barrier is complete in most of the cortices, where it restricts access of circulating antigens to developing cortical lymphocytes. Otherwise, in iuxtamedullary cortex, around cortical venules, and in the medulla,

Figure 3.1 A portrait of Morris Karnovsky.

Figure 3.2 A portrait of Elio Raviola (on the left).

barrier is not complete, allowing macromolecules and circulating antigens to penetrate from blood into thymic parenchyma. Barrier prevents antigens circulating in bloodstream from reaching thymic cortex where T-lymphocytes are formed (Janossy et al., 1981).

3.4 THE DESCRIPTION OF THE BLOOD–THYMUS BARRIER

In 1972, Raviola and Karnovsky published a work entitled "Evidence for a blood–thymus barrier using electron-opaque tracers," in which they analyzed the permeability of the vessels of the thymus in young adult mice using different electron-opaque tracers of different molecular weight, including HRP, cytochrome *c*, catalase, ferritin, and colloidal lanthanum. They demonstrated that the cortex was supplied by capillaries that have impermeable endothelial junctions, and although a small amount of tracer was transported by plasmalemmal vesicles through the capillary endothelium, this tracer was promptly sequestrated by macrophages stretched out in a continuous row along the cortical capillaries, and it did not reach the intercellular clefts between cortical lymphocytes and reticular cells (Fig. 3.3). On the contrary, the medulla contained all the leaky vessels, namely postcapillary venules and arterioles (Fig. 3.4). Across the walls of the venules, large quantities of all injected tracers escaped through the clefts between migrating

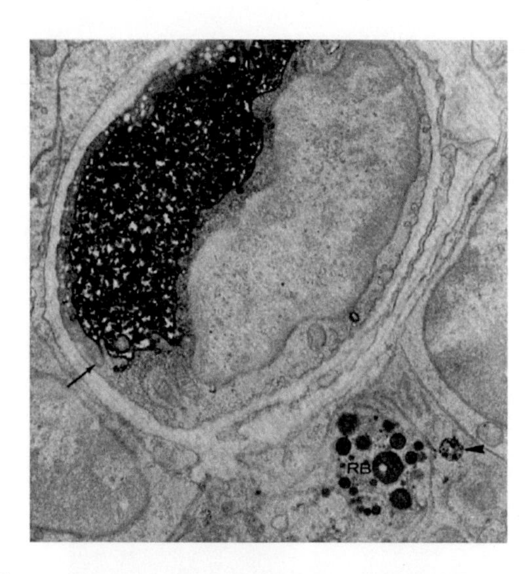

Figure 3.3 Horseradish peroxidase, 5 min after i.v. injection. Intense staining is seen in the luminal plasma of a capillary of the cortex. The abluminal end of a cleft between endothelial cells is unstained (arrow)*; also most of the plasmalemmal vesicles situated close to or opening onto the tissue front of the endothelium are unstained. The endothelial basal lamina, the adventitia, and the intercellular spaces of the surrounding cortical parenchyma are free of peroxidase. Two macrophage processes adjoining the capillary adventitia contain phagocytic vacuoles stained with reaction product (*arrowheads*). The staining of the residual bodies (*RB*) is probably nonspecific and possibly due to lipid peroxides.* Reproduced from Raviola E, Karnovsky MJ. Evidence for a blood–thymus barrier using electron-opaque tracers. J. Exp. Med. 1972;136:466–498.

lymphocytes and endothelial cells, and the arterioles showed a small number of endothelial junctions which were permeable to peroxidase but did not allow passage of tracers of higher molecular weight (Raviola and Karnovsky, 1972). The macromolecules crossing medullary venules had a limited distribution in the thymic parenchyma, because macrophages in the perivascular space ingested much of the leaked macromolecules (Naquet et al., 1999).

In their original work, Raviola and Karnovsky (1972) wrote, "The present study clearly demonstrates that in the cortex lymphocytes are protected from circulating macromolecules by a twofold mechanism: impermeability of the endothelial junctions and strategic location of the macrophages along the vessels. (...) On the contrary, circulating macromolecules have free access to the lymphocytes of the medulla. The number of medullary lymphocytes which are exposed to blood-borne substances and the concentration of these latter in the medulla are clearly related to both the rate of bulk flow macromolecules from the vascular bed and the rate of their removal from the parenchyma; in turn, the removal is probably a function of medullary macrophages,

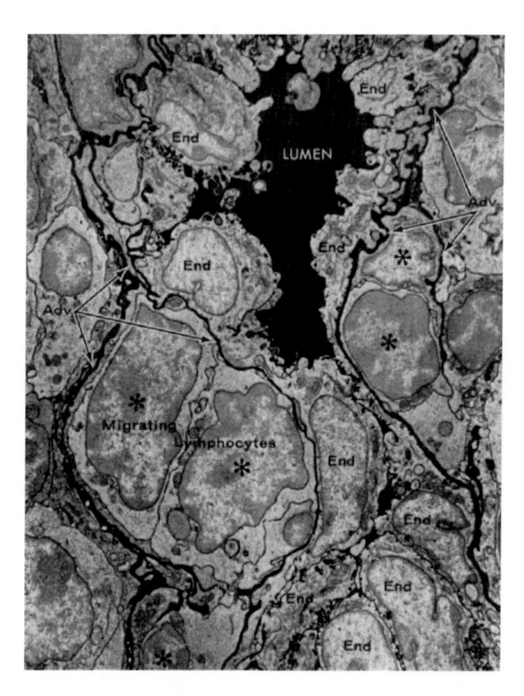

*Figure 3.4 Horseradish peroxidase, 5 min after i.v. injection. A postcapillary venule of the medulla is shown; notice its irregular endothelium (*End*) and the thin connective tissue adventitia (*Adv*), dissected by migrating lymphocytes (*asterisks*) into multiple, discontinuous layers. As a result of an impressive leakage of peroxidase, the endothelial basal lamina and the adventitia stain with the same intensity as the blood plasma.* Reproduced from Raviola E, Karnovsky MJ. Evidence for a blood–thymus barrier using electron-opaque tracers. J. Exp. Med. 1972;136:466–498.

because there is no satisfactory evidence that lymphatic vessels are present in the thymic parenchyma."

An important observation made by the Authors was the following: "While the access of circulating macromolecules to cortical lymphocytes is impeded, this may not necessarily be true for smaller substances. Small molecules, however, may enter the cortex by selective transport through the capillary endothelium and be restricted in their free movement across the vascular walls by the tight junctions which seal the endothelial clefts" (Raviola and Karnovsky, 1972).

3.5 FURTHER EVIDENCE OF THE EXISTENCE OF THE BARRIER, CONTROVERSIES AND CLINICAL RELEVANCE

The anatomical studies of Raviola and Karnovsky, using peroxidase as a tracer, indicated that the venules at the corticomedullary junction were the site of leakage for blood antigens, whereas the capillaries draining the cortex were largely impermeable. Other permeability studies have confirmed the existence of a blood–thymus barrier, allowing access to low molecular weight tracers, whereas most excluding high molecular weight particles (Kyewski, 1986; Kyewski et al., 1984).

Moreover, a perivascular system in the medulla able to trap small blood-borne molecules has been also described (Drumea-Mirancea, 2006). Entry and presentation was observed over a wide range of molecular weights, albeit with different efficiencies, and was time and dose dependent. Injection of peptides induced clonal deletion of thymocytes (Liblau et al., 1996; Murphy et al., 1990), and naturally circulating antigens were shown to induce effective negative selection of T-cell receptor transgenic T-cells in several experimental models upon presentation by thymic antigen presenting cells (Klein et al., 2001; Velazquez, 2002; Volkmann et al., 1997; Zal, 1994).

However, the existence of the blood–thymus barrier is still a matter of dispute. On the other hand, Nieuwenhuis et al. (1988) demonstrated the existence of a transcapsular pathway by which antigens may pass the blood–thymus barrier. It is clear that thymus is a target of infection and that thymic infection can be controlled if certain treatment drugs have the possibility to cross the blood–thymus barrier to directly inhibit the replication of the pathogen agent within the thymus (Savino, 2006; Nobrega et al., 2013). Infection by human immunodeficiency virus is followed by thymic involution and histological disintegration of the corticomedullary border (Napolitano et al., 2002). Similar observations have been made after parasitic infection (Pérez et al., 2012) and malnutrition (Savino et al., 2007). All these alterations are subsequently associated with increase susceptibility to autoimmune diseases and tumors (Lynch et al., 2009).

Pregnancy, as well as tumors, are associated with an involution of the thymus accompanied by a massive depletion of the cortical region (Shanker et al., 2000; Kendall and Clarke, 2000). In particular, pregnancy increases the blood–thymus barrier permeability, so that it can contribute in the development of central and peripheral tolerance. Some factors, including oncofetal antigens, prostaglandins, sex steroids like estradiol, progesterone and glucocorticoids, can impair blood–thymus barrier for foreign antigen (Martín et al., 1995; Ranga et al., 1981; Tibbetts et al., 1999). Other factors, like interleukin-2 (Cesario et al., 1991), increase the blood–thymus barrier permeability which, in turn, might be a factor involved in central and peripheral thymic tolerance (Bubanovic, 2003).

REFERENCES

Bubanovic, I.V., 2003. Failure of blood–thymus barrier as a mechanism of tumor and trophoblast escape. Med. Hypoth. 60, 315–320.

Cesario, T.C., Vaziri, N.D., Ulich, T.R., et al., 1991. Functional, biochemical, and histopathologic consequences of high-dose interleukin-2 administration in rats. J. Lab. Clin. Med. 118, 81–88.

Clark, S.L., 1963. The thymus in mice of strain 129/J, studied with the electron microscope. Am. J. Anat. 112, 1–33.

Drumea-Mirancea, M., 2006. Characterization of a conduit system containing laminin-5 in the human thymus: a potential transport system for small molecules. J. Cell. Sci. 119, 1396–1405.

Graham, R.C., Karnovsky, M.J., 1966. The early stages of absorption of injected horseradish peroxidase in the proximal tubules of mouse kidney: ultrastructural cytochemistry by a new technique. J. Histochem. Cytochem. 14, 291–302.

Janossy, G., Thomas, J.A., Goldstein, G., et al., 1981. The Human Thymic Microenvironment. Novartis Found Symp. Wiley-Blackwell, Chichester, UK, pp. 193–214.

Karnovsky, M.J., 1967. The ultrastructural basis of capillary permeability studied with peroxidase as a tracer. J. Cell. Biol. 35, 213–236.

Karnovsky, M.J., 2006. A pathologist's odyssey. Annu. Rev. Pathol. Mech. Dis. 1, 1–22.

Kendall, M.D., Clarke, A.G., 2000. The thymus in the mouse changes its activity during pregnancy: a study of the microenvironment. J. Anat. 197, 393–411.

Klein, L., Roettinger, B., Kyewski, B., 2001. Sampling of complementing self-antigen pools by thymic stromal cells maximizes the scope of central T cell tolerance. Eur. J. Immunol. 31, 2476–2486.

Kyewski, B.A., 1986. Intrathymic presentation of circulating non-MHC antigens by medullary dendritic cells. An antigen-dependent microenvironment for T cell differentiation. J. Exp. Med. 163, 231–246.

Kyewski, B.A., Fathman, C.G., Kaplan, H.S., 1984. Intrathymic presentation of circulating non-major histocompatibility complex antigens. Nature 308, 196–199.

Liblau, R.S., Tisch, R., Shokat, K., et al., 1996. Intravenous injection of soluble antigen induces thymic and peripheral T-cells apoptosis. Proc. Nat. Acad. Sci. 93, 3031–3036.

Lynch, H.E., Goldberg, G.L., Chidgey, A., et al., 2009. Thymic involution and immune reconstitution. Trends Immunol. 30, 366–373.

Marshall, A.H., White, R.G., 1961. The immunological reactivity of the thymus. Br. J. Exp. Pathol. 42, 379–385.

Martín, A., Casares, F., Alonso, L., et al., 1995. Changes in the blood thymus barrier of adult rats after estradiol-treatment. Immunobiol 192, 231–248.

Murphy, K., Heimberger, A., Loh, D., 1990. Induction by antigen of intrathymic apoptosis of CD4 + CD8 + TCRlo thymocytes in vivo. Science 250, 1720–1723.

Napolitano, L.A., Lo, J.C., Gotway, M.B., et al., 2002. Increased thymic mass and circulating naive CD4 T cells in HIV 1-infected adults treated with growth hormone. AIDS. 16, 1103–1111.

Naquet, P., Naspetti, M., Boyd, R., 1999. Development, organization and function of the thymic medulla in normal, immunodeficient or autoimmune mice. Sem. Immunol. 11, 47–55.

Nieuwenhuis, P., Stet, R.J., Wagenaar, J.P., et al., 1988. The transcapsular route: a new way for (self-) antigens to by-pass the blood–thymus barrier. Immunol. Today. 9, 372–375.

Nobrega, C., Nunes-Alves, C., Cerqueira-Rodrigues, B., et al., 2013. T cells home to the thymus and control infection. J. Immunol. 190, 1646–1658.

Pérez, A.R., Berbert, L.R., Lepletier, A., et al., 2012. TNF-α is involved in the abnormal thymocyte migration during experimental *Trypanosoma cruzi* infection and favors the export of immature cells. PLoS ONE. 7, e34360.

Ranga, V., Ispas, A.T., Chirulescu, A.R.M., 1981. Elements of structure and infrastructure of the blood–thymus barrier in ACTH involuted thymus. Acta Anat. 111, 177–189.

Raviola, E., Karnovsky, M.J., 1972. Evidence for a blood–thymus barrier using electron-opaque tracers. J. Exp. Med. 136, 466–498.

Reese, T.S., Karnovsky, M.J., 1967. Fine structural localization of a blood–brain barrier to exogenous peroxidase. J. Cell. Biol. 34, 207–217.

Savino, W., 2006. The thymus is a common target organ in infectious diseases. PLoS Pathog. 2, e62.

Savino, W., Dardenne, M., Velloso, L.A., et al., 2007. The thymus is a common target in malnutrition and infection. Br. J. Nutr. 98 (Suppl 1), S11–S16.

Shanker, A., Singh, S.M., Sodhi, A., 2000. Ascitic growth of a spontaneous transplantable T cell lymphoma induces thymic involution. 2. Induction of apoptosis in thymocytes. Neoplasma. 47, 90–95.

Straus, W., 1957. Segregation of an intravenously injected protein by "Droplets" of the cells of rat kidneys. J. Cell. Biol. 3, 1037–1040.

Tibbetts, T.A., DeMayo, F., Rich, S., et al., 1999. Progesterone receptors in the thymus are required for thymic involution during pregnancy and for normal fertility. Proc. Nat. Acad. Sci. U.S.A. 96, 12021–12026.

Velazquez, C., 2002. Chemical identification of a low abundance lysozyme peptide family bound to I-Ak histocompatibility molecules. J. Biol. Chem. 277, 42514–42522.

Volkmann, A., Zal, T., Stockinger, B., 1997. Antigen presenting cells in the thymus that can negatively select MHC class II-restricted T cells recognizing a circulating self antigen. Immunol. Lett. 56, 87–88.

Weiss, L., 1963. Electron microscopic observations on the vascular barrier in the cortex of the thymus of the mouse. Anat. Rec. 145, 413–437.

Zal, T., 1994. Mechanisms of tolerance induction in major histocompatibility complex class II-restricted T cells specific for a blood-borne self-antigen. J. Exp. Med. 180, 2089–2099.

FURTHER READING

Schneeberger-Keeley, E.E., 1968. The ultrastructural basis of alveolar–capillary membrane permeability to peroxidase used as a tracer. J. Cell. Biol. 37, 781–793.

CHAPTER 4

Max D. Cooper and the Delineation of Two Lymphoid Lineages in the Adaptive Immune System

4.1 BIOGRAPHIC NOTES

Max D. Cooper (Fig. 4.1) attended medical school at "Tulane University" and received doctoral degree in 1957. After a year spent at the University of San Francisco, Cooper worked to establish the dual nature of the immune system with Robert A. Good as postdoctoral fellow and Assistant Professor at the University of Minnesota from 1963 until 1967. He was the appointed Associate Professor at the University of Alabama at Birmingham where he remained for the next 41 years, as Professor of Immunology at the Departments of Pediatrics, Medicine, Microbiology, and Pathology. During his career he has been a visiting scientist in the Tumor Immunology Unit at the University College London and at Insititut d'Embryologie at Institut Pasteur, Paris.

In 1974, while on sabbatical at the University College in London, he worked with Martin Raff and John Owen to identify the bone marrow and fetal liver precursors of B cells. He has received the Sandoz Prize in Immunology, American College of Physicians Science Award, American Association of Immunologists (AAI) Lifetime Achievement Award, Avery-Landsteiner Prize, and the Robert Koch Award. He served as president of the Clinical Immunology Society (1994) and of the AAI (1988).

4.2 THE ROLE OF THE BURSA OF FABRICIUS AND OF THE THYMUS IN THE DEFINITION OF TWO PATTERNS OF LYMPHOCYTE LINEAGES

The bursa of Fabricius and the thymus are the central lymphoid organs in the chicken essential for the ontogenetic development of the adaptive immunity. In the chicken embryo, the thymus is the first lymphoid organ to develop. The epithelial component is evident before the

Milestones in Immunology. DOI: http://dx.doi.org/10.1016/B978-0-12-811313-4.00004-8

Figure 4.1 A portrait of Max D. Cooper.

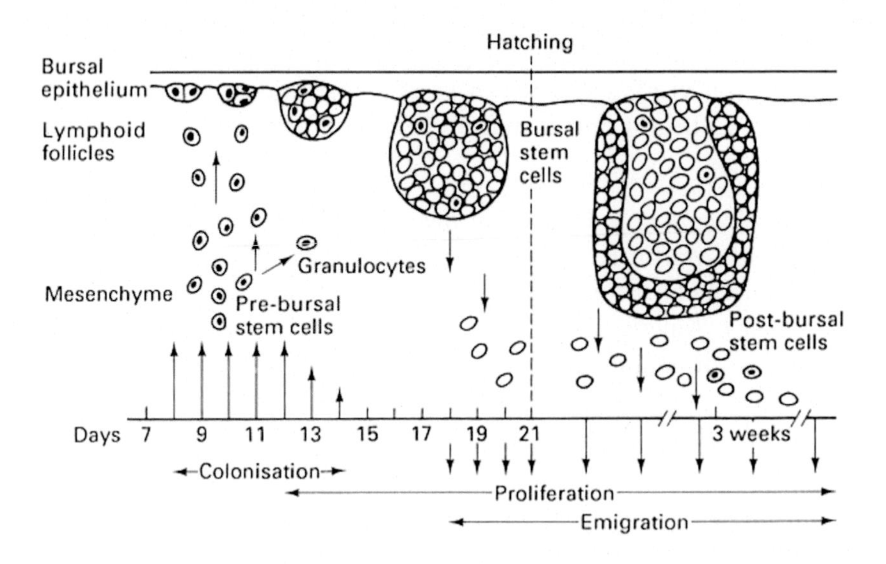

Figure 4.2 Time-course of the morphogenesis of lymphoid follicles in the bursa of Fabricius.

9th day of incubation, and the thymus is a fully developed lymphoid organ by the 12th day of incubation. Between the 12th and the 14th day, budding of the epithelial fold of the bursa is observed, and on the 14th day the lymphoid structures begin to develop by direct transformation of epithelial cells to lymphoid cells (Fig. 4.2).

Glick et al. (1956) demonstrated that the bursa plays an important role in antibody production, showing that antibody responses are suppressed in the majority of bursectomized chickens. As Cooper (2015) remembered: "The report of these findings was rejected by a mainstream journal because it was not considered of general interest and its publication in 'Poultry Science' in 1956 was unnoticed at the time by immunologists. Thus although all of these observations were indicative of a distinction between cellular and humoral immunity, an integrated interpretation of the results was obscured at the time by their derivation from experiments in disparate species."

Miller (1961) discovered the role of thymus-derived cells in cellular immunity. Metcalf (1960) studied the peripheral blood lymphocyte levels and histology of the lymphoid tissues in mice thymectomized between 4 and 6 weeks of age and demonstrated that there was a slow progressive fall in circulating lymphocytes to a maximum of 30% to 40% below normal values.

Work in several laboratories from 1962 to 1964 suggested that the bursa and the thymus probably did have different functions. The functional dissociation within/of the chicken immune system was first suggested by Szenberg and Warner (1962). Observations on the changes in the lymphoid organs after bursectomy and thymectomy in chickens have indicated the possible existence of two almost completely separate lymphocytopoietic systems (Fig. 4.3). Carey and Warner (1964) showed that hormonally bursectomy significantly reduced circulating immunoglobulin levels and Ortega and Der (1964) were able to produce consistent 7S agammaglobulinemia by complete surgical bursectomy on the day of hatching. Isakovic and Jankovic (1964) suggested that responses of the delayed hypersensitivity type and the small lymphocyte development in the tissues were probably dependent on the thymus in the chicken, as they were in mice and rats.

The earliest scientific contribution of Max Cooper defined the basis of adaptive immunity. In 1964, as postdoctoral fellow in the laboratory of Robert Good, he discovered the dual origin of lymphoid cells in the chicken. Earlier removal of the thymus and of the bursa was needed to clarify their respective roles in immune system development.

As Good (1995) has written: "Max Cooper, then an allergist–pediatrician, had just come to our laboratory to begin his immunology

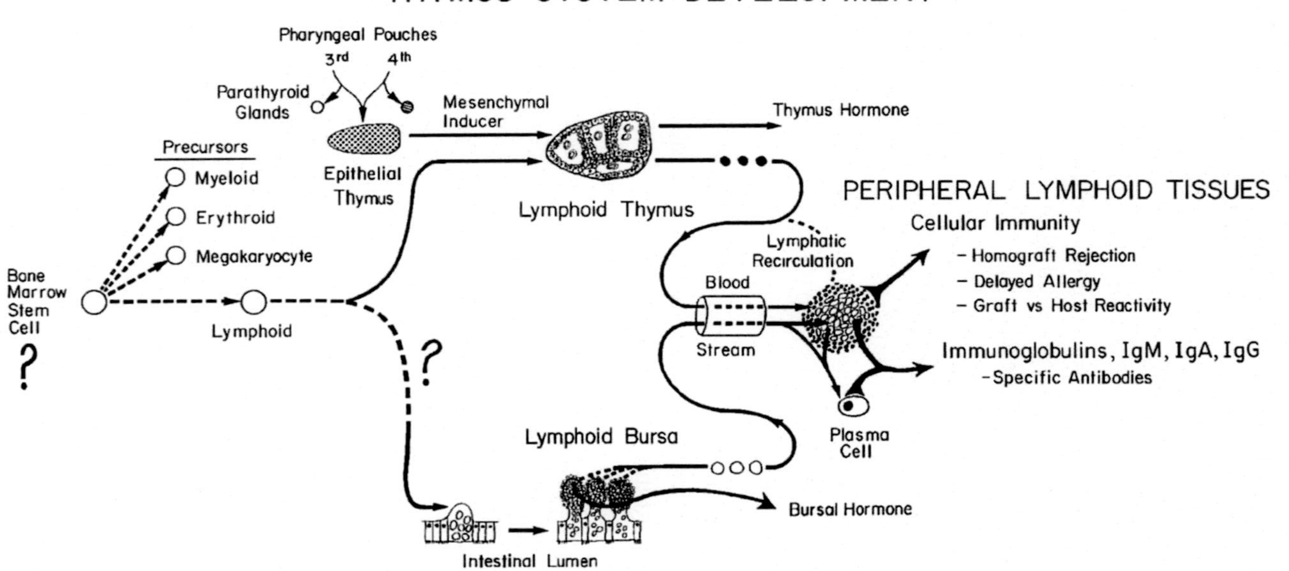

Figure 4.3 An original model of Max D. Cooper concerning the different development of thymus and bursal systems. Reproduced from Cooper M, Gabrielsen A, Good R. Central peripheral lymphoid tissues in immunologic processes and human disease. In: Mayerson H, editor. Lymph and the Lymphatic System. Springfield: Charles C. Thomas, 1968: p. 276–305.

post doctorate fellowship training. He decided to revisit in laboratory studies the roles played by the thymus and bursa on lymphoid development in the chicken. Removal of the thymus from x-irradiated, newly hatched chickens, removal of the bursa of Fabricius from similarly sublethally irradiated newly hatched chickens, or removal of both thymus and bursa from such newly hatched chickens each produced very different results. Removal of the thymus so early in life prevented the development of lymphocytes in the blood and in the dense aggregates of lymphocytes in the white pulp of chicken spleen, leaving the germinal center and plasma cell development impressively intact."

On the other hand, in the mind of Cooper (2010): "The plan was to compare the immunological status of the different experimental groups, after they recovered from the effects of surgery and irradiation." He specified, "I devised an alternative strategy that would combine posthatching thymectomy or bursectomy together with whole body irradiation to destroy cells that might have seeded earlier from the thymus and bursa or that could have been influenced by postulated thymic and bursal humoral factors. In these experiments I removed either the thymus or the bursa, then subjected the newly hatched chicks to near lethal irradiation and waited several weeks until they and their irradiated controls recovered from the irradiated effects." (Cooper, 2002).

Bursectomized and irradiated birds were characterized by the absence of germinal centers, plasma cells, and circulating immunoglobulins, or the ability to make antibodies in response to immunization. This evidence clearly shows that the bursa provides a unique microenvironment for the proliferation and differentiation of B cells (Ratcliffe, 2006). On the other hand, thymectomized and irradiated animals were deficient in lymphocytes that mediated inflammatory responses, as assessed by skin graft rejection, delayed-type hypersensitivity, and graft versus host reaction. Birds subjected to combined thymectomy and bursectomy and irradiation had severe cellular and humoral immune defects (Cooper et al., 1965, 1966a).

As Good (1995) has precised: "The thymus, we had found, was related to the development of a population of small lymphocytes, which we called thymus-dependent cells and were later named T cells by Ivan Roitt. They were located especially in blood, in splenic white pulp aggregates, and in specific regions of the small lymph nodes and gastrointestinal lymphoid aggregates in the chicken." Cooper (2002) has concluded that: "Our model of the development of separate lymphocyte lineages provide a reliable operational map of these two differentiation pathways (...) This model radically changes our perspective of lymphocyte differentiation defects in patients with primary or secondary immunodeficiency diseases." Moreover, "The B cell and T cell lineage model raised many basic questions. What is the source of B cells in mammals? If B cells use their antibodies as antigen receptors, what do T use to recognize antigen given their inability to make antibodies? How do T cells cooperate with B cells to facilitate antibody response?" (Cooper, 2015).

4.3 THE IMMUNOGLOBULIN ISOTYPE SWITCHING

Cooper et al. (1965) by means of immunofluorescence analysis with antisera specific for mu and gamma chains demonstrated that the bursa is the first site where cells produce mu chains, and probably immunoglobulin M (IgM). The first surface IgM positive cells are detected from day 12 of incubation and at hatching more than 90% of bursal cells are mature B cells. Later, the bursa it is the first site where gamma chains and probably immunoglobulin G (IgG) are produced. Injection of a specific antibody against the mu chain into the developing chick embryo at the moment of appearance of IgM-staining cells in the bursa prevents the development of both IgM- and IgG-producing cells in chicken. Moreover, when specific goat antiserum against mu chains was tagged with fluorescin isothiocianate and specific antiserum against gamma chains was tagged with rhodamine, it was clear that the bursa was the first site to develop both IgM- and IgG-producing lymphocytes. Bursectomy at 17 to 19 days of embryonic development prevented development of a population of IgG-producing plasma cells. Chicken bursectomized and irradiated at hatching fail to develop either IgM or IgG and cannot make antibodies. Such animals develop into agammaglobulinemic animals lacking all cells of the bursa-dependent lines, and they do not have lymphocytes whose receptor immunoglobulins can be identified. The infusion of autologous bursal lymphocytes restored germinal center development, plasma cell generation, and immunoglobulin production in bursectomized and irradiated chicks (Cooper et al., 1966b).

As Cooper (2015) pointed out: "These studies provided insights into the genetic basis for the orderly expression of clonally diverse IgM heavy and then light chain genes, the subsequent switch in constant region genes to enable B cells to express different antibody classes and the concomitant antigen-mediated selection of somatic mutations."

4.4 MAMMALIAN "BURSA-EQUIVALENT" ORGANS AND THE ROLE OF LIVER AND BONE MARROW IN LYMPHOPOIESIS

Cooper et al. (1966c) formulated the hypothesis, "gut-associated lymphoepithelial tissues (GALT) serve as the mammalian bursa equivalent in supplying B cells to the rest of the body." "The tonsil was my first candidates for the bursa-equivalent, but their removal in newborn rabbits had no effect on antibody production" (Cooper, 2010). Sutherland et al. (1964) had shown that neonatal appendectomy impaired antibody response in rabbits. Cooper et al. (1966b) decided on the strategy of neonatal appendectomy followed by Peyer's patch removal in combination with whole body irradiation to destroy preexisting lymphocytes. They found that these rabbits had immunological defects comparable to those observed in older chickens subjected to bursectomy and irradiation, whereas the ability to reject skin allografts was maintained. Peyer's patches may be special sites where antigen-driven proliferation can lead to great expansion of a B-cell population and to a switching of capacity to produce one kind of immunoglobulin, IgM, IgG, or immunoglobulin A (IgA).

In 1974, in collaboration with John Owen in London, Cooper performed experiments to verify that B cells were produced in hematopoietic tissues. They demonstrated that when mouse fetal liver was placed in culture before the appearance of B cells, B cells were indeed generated (Owen et al., 1974). In a further study, they demonstrated that B cells were also generated ex vivo in mouse fetal long bones (Owen et al., 1976). In subsequent experiments, Owen et al. (1977) showed that large pre-B cells undergo proliferation before giving rise to small resting pre-B cells that, in turn, differentiate to become B cells.

Overall, these findings suggested, "mammalian B-cell generation is a multifocal process that shifts from one hematopoietic environment to another during development, and B lymphopoiesis was later shown to continue throughout life in the bone marrow" (Cooper, 2002). Now, it is well established that mature B cells originate primarily in the bone marrow in adult mammals, a process which is antigen independent, from hematopoietic stem cells.

4.5 CLINICAL CORRELATES

The model of the development of separate lymphocyte lineages proposed by Cooper (2002) "radically changed our perspective of lymphocyte differentiation defects in patients with primary or secondary immunodeficiency diseases." As Good has emphasized: "Our experiments with irradiated bursectomized chicken combined with our similarly extensive histopathologic studies of patients with primary immunodeficiency diseases provided a clear perspective for understanding several of these immunodeficiency diseases and also the entire lymphoid system in both chickens and humans" (Cooper et al., 1968).

Perhaps, the earliest immunologic deficiency to be attributed to a failure of thymic development was the classic Swiss lymphopenic form of agammaglobulinemia. Indeed, in 1950 this disease was well described as an almost complete failure of lymphoid development in early infancy (Glanzmann and Rinker, 1950). Since 1950, many reports of extreme thymus-system deficiencies of infancy and early childhood appeared.

Bursectomized chickens are similar to patients with Bruton's X-linked agammaglobulinemia, and thymectomized chickens are similar to patients with Di George syndrome, whereas bursectomized and thymectomized in the newly hatched period chickens are similar to patients with severe combined immunodeficiency disease (Peterson et al., 1965). The absence of germinal centers, plasma cells, and antibodies in patients with Bruton's X-linked agammaglobulinemia, who had a normal thymus and intact cell-mediated immunity, could be explained by an arrest in differentiation of the immunoglobulin-producing lineage of lymphocytes.

As Good has pointed: "Immediately following Cooper's presentation, Di George from St. Christopher's Hospital in Philadelphia described his observations on children born without thymus or parathyroid glands who often exhibited these defects together with congenital cardiac defects and especially with abnormalities of the outflow tracts of the heart" (Cooper et al., 1968).

In the Di George syndrome, parathyroid and thymus development are both lacking, but the immunoglobulin-producing system is relatively intact. Patients with Di George syndrome develop all immunoglobulins well and form antibody to many antigens very well, but fail to produce cellular immunity. They possess germinal centers and plasma cells, Peyer's patches, and tonsils. By contrast, the lymphocyte population in the deep cortical or paracortical thymus dependent zones are deficient. Although this may have been the initial description of patients affected with Di George syndrome, it is well established by now that these patients may also present variable humoral defects ranging from selective IgA deficiency to hypogammaglobulinemia of all classes (Patel et al., 2012).

Cooper et al. (1971) showed that boys with X-linked agammaglobulinemia have pre-B cells in their bone marrow, but very few B cells, findings indicative of an early arrest in B cell lineage differentiation.

Parallel to immunodeficiency, "we began to analyze lymphoid malignancies in patients together with our hematology-oncology colleagues" (Cooper, 2010). Lymphoid malignancies could be classified according to whether the neoplastic lymphoid differentiation began in the thymus or the mammalian bursa equivalent. The immunologic deficiencies associated with lymphoproliferative disorders have for some time sorted into two groups. Before the defects were defined with the advent of serum electrophoresis and more sophisticated immunological assay methods, it had long been known that the major infectious threats to such patients tended to "cluster" into two groups, with tuberculosis and mycotic infections on the one hand, and infections with extracellular pyogenic pathogens on the other.

The majority of malignancy of the lymphoid tissue occurring in mice involve the thymus-dependent small lymphocytes. In many mouse leukemias, the malignancy process begins in the thymus. In some, the tumors progress no further and the animals die early of pulmonary insufficiency consequent to a large thymoma. When in these malignancy tumor cells are present elsewhere, as in the spleen and lymph nodes, they involve predominantly the small lymphocyte population.

Vogler et al. (1978) found that acute lymphocytic leukemia (ALL) of childhood represent pre-B cell tumors. In ALL, blockade of lymphoid cell differentiation leads to persistent proliferation, defective cell death, and accumulation of leukemic lymphoblasts in tissues. These patients have normal immunoglobulins, intact capacity for antibody production, and normal cell mediated immune responses including development and expression of delayed hypersensitivity and homograft reaction. In these patients, lymphocyte responses to phytohemagglutinin was normal

(Astaldi et al., 1965). These patients, however, had a high frequency of infections particularly to low grade pyogenic pathogens like staphylococci, pseudomonas, and certain gram-negative bacteria, which may be attributed to the marked deficiency of granulocytes.

4.6 CONCLUDING REMARKS

The role of the thymus and of the bursa of Fabricius in the development of immunologic competence both before and after birth has placed a new focus on the ontogeny of the lymphoid tissues. Speculation on the reason for immunological failure following neonatal bursectomy and thymectomy has indicated the thymus as a source of cells or humoral factors essential to normal lymphoid development and immunologic maturation. Both clinical and experimental data are accumulated in support of the thesis that absence of the bursa and/or thymus during the period of development of the peripheral lymphoid tissues results in both abnormal lymphoid development and immunologic deficiency.

REFERENCES

Astaldi, G., Sauli, S., Airo, R., et al., 1965. Effect of phytohemagglutinin on lymphocytes from different leukemias. Texas Rep. Biol. Med. 23, 569.

Carey, J., Warner, N.L., 1964. Gamma-globulin synthesis in hormonally bursectomized chickens. Nature. 203, 198.

Cooper, M., Gabrielsen, A., Good, R., 1968. Central peripheral lymphoid tissues in immunologic processes and human disease. In: Mayerson, H. (Ed.), Lymph and the Lymphatic System. Charles C. Thomas, Springfield, pp. 276–305.

Cooper, M.D., 2002. Exploring lymphocyte differentiation pathways. Immunol. Rev. 185, 175–185.

Cooper, M.D., 2010. A life of adventure in immunobiology. Annu. Rev. Immunol. 28, 1–19.

Cooper, M.D., 2015. The early history of B cells. Nat. Rev. Immunol. 15, 191–197.

Cooper, M.D., Peterson, R.D., Good, R.A., 1965. Delineation of the thymic and bursal lymphoid systems in the chicken. Nature 205, 143–146.

Cooper, M.D., Raymond, D.A., Peterson, R.D., et al., 1966a. The functions of the thymus system and the bursa system in the chicken. J. Exp. Med. 123, 75–102.

Cooper, M.D., Schwartz, M.L., Good, R.A., 1966b. Restoration of gamma globulin production in agammaglobulinemic chickens. Science 151, 471–473.

Cooper, M.D., Perey, D.Y., McKneally, M.F., et al., 1966c. A mammalian equivalent of the avian bursa of Fabricius. Lancet 1, 1388–1391.

Cooper, M.D., Lawton, A.R., Bockman, D.E., 1971. Agammaglobulinaemia with B lymphocytes. Specific defect of plasma-cell differentiation. Lancet 2, 791–794.

Glanzmann, E., Rinker, P., 1950. Esssentielle Lymphocytophthise. Ein neues Krankheitsbild aus der Säuglingspathologie. Ann. Pediatr. 175, 1.

Glick, C., Chang, T., Jaap, R., 1956. The bursa of Fabricius and antibody production. Poultry Sci. 35, 224–234.

Good, R.A., 1995. Organization and development of the immune system. Relation to its reconstruction. Ann. N.Y. Acad. Sci. 770, 8–33.

Isakovic, K., Jankovic, D.B., 1964. Role of the thymus and the bursa of Fabricius in immune reactions in the chickens. II. Cellular changes in lymphoid tissues of thymectomized, bursectomized and normal chickens in the course of first antibody response. Int. Arch. Allergy 24, 296.

Miller, J.F., 1961. Immunological function of the thymus. Lancet. 2, 748–749.

Metcalf, D., 1960. The effect of thymectomy on the lymphoid tissues of the mouse. Br. J. Haematol. 6, 324–333.

Ortega, L.G., Der, B.K., 1964. Studies of agammaglobulinemia induced by ablation of the bursa of Fabricius. Fed. Proc. 23, 546.

Owen, J.J., Cooper, M.D., Raff, M.C., 1974. In vitro generation of B lymphocytes in mouse foetal liver, a mammalian 'bursa equivalent'. Nature 249, 361–363.

Owen, J.J., Raff, M.C., Cooper, M.D., 1976. Studies on the generation of B lymphocytes in the mouse embryo. Eur. J. Immunol. 5, 468–473.

Owen, J.J., Wright, D.E., Habu, S., et al., 1977. Studies on the generation of B lymphocytes in fetal liver and bone marrow. J. Immunol. 118, 2067–2072.

Patel, K., Akhter, J., Kobrynski, L., et al., 2012. Immunoglobulin deficiencies: the B-lymphocyte side of Di George syndrome. J. Pediatr. 161, 950–953.

Peterson, R.D., Cooper, M.D., Good, R.A., 1965. The pathogenesis of immunologic deficiency diseases. Am. J. Med. 38, 579–604.

Ratcliffe, M.J.H., 2006. Antibodies, immunoglobulin genes and the bursa of Fabricius in chicken B cell development. Dev. Comp. Immunol. 30, 101–118.

Sutherland, D.E., Archer, O.K., Good, R.A., 1964. Role of the appendix in development of immunologic capacity. Proc. Soc. Exp. Biol. Med. 115, 673–676.

Szenberg, A., Warner, N., 1962. Dissociation of immunological responsiveness in fowls with hormonally development of lymphoid tissues. Nature 194, 146.

Vogler, L.B., Crist, W.M., Bockman, D.E., et al., 1978. Pre-B-cell leukemia. A new phenotype of childhood lymphoblastic leukemia. N. Engl. J. Med. 298, 872–878.

CHAPTER 5

Edelman's View on the Discovery of Antibodies

5.1 BIOGRAPHIC NOTE

Gerald M. Edelman (Fig. 5.1) died on May 17, 2014, aged 84 years, in La Jolla, California. Born in New York City, on July 1, 1929, he studied medicine at the University of Pennsylvania, where he received medical degree in 1954. From 1955 to 1957, Edelman served in the Army Medical Corps in Paris, after which he began working to the structure of antibodies, when he joined as graduate student the laboratory of Henry Kunkel in 1958 at the "Rockefeller University" in New York, obtaining his doctorate in 1960. As Edelman (2004a) pointed out: "In 1958, I was fascinated by the specificity of antigen recognition by antibodies. At the same time I was frustrated as a chemist by the heterogeneity of the γ-globulin fractions containing these antibodies. Free boundary electrophoresis by Arne Tiselius and Elvin Kabat revealed a stark contrast between the distribution of net charge of these proteins as compared with that of other serum proteins. I was driven by the hope of resolving this heterogeneity and had the dream, naive as it was at the time, that by doing the primary structure of antibody molecules, the basis of their specificity would be revealed."

Edelman's focus on the structure of antibodies led to the 1972 Nobel Prize in Physiology or Medicine along with Rodney R. Porter (1917–86) (Fig. 5.2) "for their discoveries concerning the chemical structure of antibodies." Their advance "incited a fervent research activity the whole world over, in all fields of immunological science, yielding results of practical value for clinical diagnostics and therapy," the Nobel committee said. The award recognized Edelman's work on the descriptions of the heavy (H) and light (L) polypeptide chains, and Porter's on the distinct binding domain (antigen-binding or Fab fragment) of antibodies.

Gerald Edelman's career has been eclectic and highly successful. After the prize, Edelman's group expanded its interests in several areas, including carbohydrate-binding proteins on the surface

Milestones in Immunology. DOI: http://dx.doi.org/10.1016/B978-0-12-811313-4.00005-X

Figure 5.1 A port trait of Gerald M. Edelman.

Figure 5.2 A port trait of Rodney R. Porter.

membrane of cells and tissue morphogenesis. He found a class of proteins, called cell adhesion molecules, thought to play an important role in embryonic development.

In the late 1970s, Edelman turned his attention to the brain, and by 1991, he was director of the "Neuroscience Institute" at "Rockefeller University." In 1993, he moved the institute to the La Jolla neighborhood of San Diego. From 1995, the institute was part of the Scripps Research Institute campus; it moved to another location in La Jolla in 2012.

Edelman formulated a theory to explain the development and organization of higher brain functions in terms of a process known as neuronal group selection. This theory was presented in his 1987 volume Neural Darwinism. A new, biologically based theory of consciousness extending the theory of neuronal group selection is presented in his 1989 volume The Remembered Present. A subsequent book, Bright Air, Brilliant Fire, published in 1992, continues to explore the implications of neuronal group selection and neural evolution for a modern understanding of the mind and the brain.

5.2 THE DISCOVERY OF THE STRUCTURE OF IMMUNE GLOBULINS

In the 1950s, immunology was largely a subdiscipline of microbiology with a strong emphasis on serology and vaccines. The prevailing theories were instructional: accordingly, the antigen or hapten served as a template for the folding of the antibody-combining site (Pauling, 1940) (Fig. 5.3). The "instructional" theory, as it was proposed by Pauling, arose from the universally accepted concept that the antibody repertoire had to be transcendental to protect the animal. As the number of genes had to be lower than the number of antigens that are recognized, the only solution seemed to be the "instructional" theory. Accordingly, the antibody was synthesized or folded, in specific ways in spatial contact with the antigen, which acted as a template. In the opinion of Edelman (2004a): "Pauling looked only at the chemical level. He ignored the fact that the body did not produce antibodies to its own antigens, a fact difficult to account for simply by instructional folding."

Within the period from 1959 to 1969, instructional theories were abandoned in favor of selection theories (Jerne, 1955; Burnet, 1959),

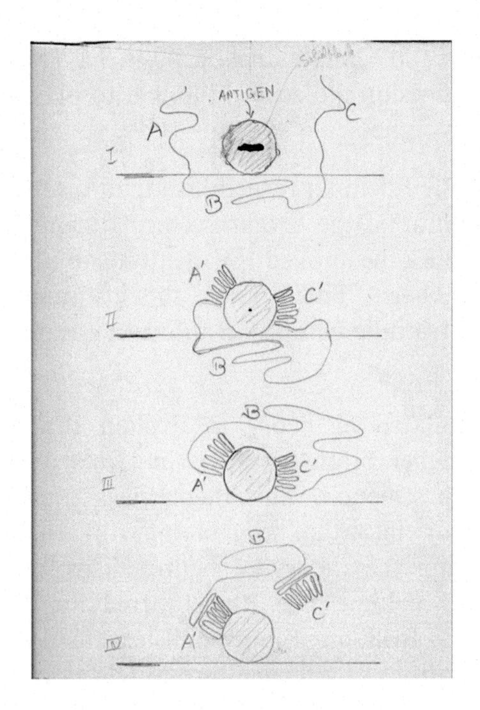

Figure 5.3 Drawing of antigen and antibodies by Linus Pauling.

that took their inspiration from Darwinian two-step variation and selection, leading to alterations in population structure. The various immune globulins were synthesized prior to antigen exposure and made up a repertoire of variant proteins.

In his paper entitled "The natural selection theory of antibody formation," Jerne wrote, "Among the comparatively small number, perhaps a few thousand, of antigen–antibody systems investigated, cross-reactions are by no means rare, suggesting that the number of specific configurations which globulin molecule can exhibit is large but limited. Since normal mammalian serum contains more than 10^{17} globulin molecules per millimeter, these may include a million 10^{11} fractions of different specificity. This would seem an amply sufficient number."

Burnet suggested that an antigen might induce modifications of those enzymes involved in globulin synthesis so that a protein with the required specificity might be formed. This hypothesis concerned the manner in which the body normally failed to make antibodies to its own components, the "self-marker" concept, according to which the distinction between the "self" and the "not-self" is based on a limited number of recognizable components in each cell, the combination of which constitute the specific "self-pattern" of the organism. The basis of clonal selection theory is that the specific capacity of a cell to react immunologically is conferred on by genetic processes and not by the intrusion of a pattern from the antigen. The clonal selection theory advanced the concept that antibodies were natural globulins that possessed an affinity for antigens and are selected from a large group of preexisting globulins. In the opinion of Edelman (2004a): "Burnet was insufficiently respectful of the biochemical rules, the syntax that would ultimately reveal in detail the origins of antibody diversity and specificity."

Edelman and Porter decided to approach the problem of antibodies structure by splitting. Porter's experiments focused on a specific antibody from the gamma globulin fraction of rabbit serum called immunoglobulin G or IgG. Porter (1959) published a report in which he used the enzyme papain to cleave the antibody molecule into three pieces of about 50,000 Da, corresponding to the two Fab (antigen-binding) and constant Fc (crystallizable) fragments (Figs. 5.4 and 5.5). Papain hydrolyzed peptide bonds in IgG and produced three fragments

Figure 5.4 The cover of "Nature Immunology" dedicated to Burnet's clonal selection theory.

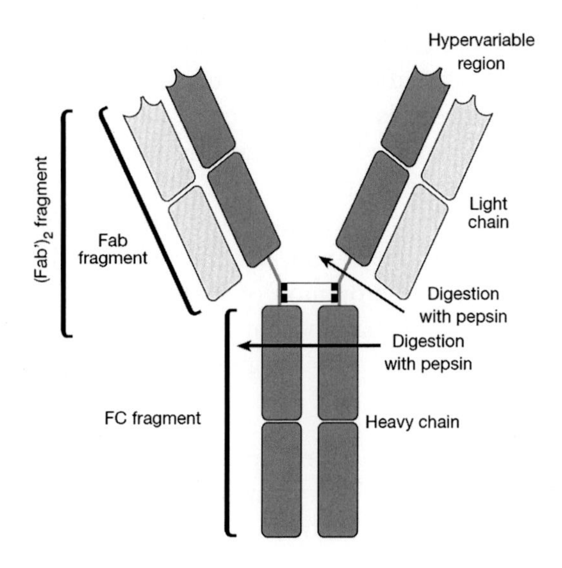

Figure 5.5 Monomeric unit structure of immune globulins. Enzymatic digestion with papain produces two fragments that bind the antigen (Fab) and one that crystallizes (Fc). Treatment with pepsin produces one large fragment (Fab')2 and smaller ones (pepsin peptides).

(I, II, and III). The three fragments had similar molecular weights (50 kDa) but different charges.

The crystals of the Fc fragments coming from antibodies with different specificities were practically homogeneous. On the other hand, the lack of capacity of Fab fractions to form crystals correlated the antigenic specificity with the structural heterogeneity and differences in its amino acid sequence.

In the same year, Edelman (1959) showed that reduction of the disulfide bonds of antibodies in the presence of denaturizing agents led to dissociation of the molecule into smaller pieces, now known to be the light (L) and heavy (H) chains. Because the molecular weight of the original IgG molecule is 150 kDa, he concluded that the IgG molecule consisted of two heavy and two light chains linked by disulfide bonds and noncovalent interactions. A Y-shaped configuration was proposed and then confirmed through electron microscopy and x-ray diffraction study. Thereafter, two antigenic types of light chains, denominated κ and λ chains were described (Fig. 5.6).

At that time, Sanger (Stretten, 2002) had completed the sequencing of insulin and Stein and Moore (Hirs et al., 1960) were sequencing

Figure 5.6 Gerald M. Edelman at Rockefeller University in 1972, in front of a gamma globulin model.

ribonuclease. As Edelman (2004b) wrote: "I had the thought that if I broke the disulfide bonds in gamma globulin, I could look in the analytical centrifuge to see what happened to the molecule (...) It looked like the treated molecule hardly migrated at all compared to the original one (...) I come to the conclusion that the molecule must have composed of more than one polypeptide chain linked by disulfide bonds (...) My paper, by the way, was almost a phantom. It was a single page with one introductory sentence, a one sentence conclusion and an equation in the middle. The conclusion was that the gamma globulin molecule consists of multiple subunits of polypeptide chains."

Following upon demonstration by Hilschmann and Craig (1965) that light chains were composed of variable and constant regions, Edelman et al. (1969) established the complete amino acid sequence of a myeloma protein. He picked one IgG myeloma protein that had obtained in over 200 g-amounts from a patient who had been exchange-transfused because his myeloma protein was caused viscosity problems, and completed the sequence of the whole molecule (Edelman et al., 1969). In the meantime, Press and Hogg (1969) had most of their heavy chain and comparison of the data from both laboratories with those on other heavy chains confirmed the hypothesis that there were V and C regions from these chains.

Edelman (1970) proposed the domain hypothesis predicting that the amino acid sequence in each region would fold into a compact domain with a single disulfide bond in its interior. However, this hypothesis proposed that each pair of domains across a symmetry axis (L–H or

H–H) would carry out a distinct function and, finally, domains were presumed to have arisen by gene duplication of an ancestral gene coding for a single domain.

5.3 THE DISCOVERY OF L-CHAIN POLYPEPTIDE IN THE URINE OF MULTIPLE MYELOMA PATIENTS

The application of protein sequencing methodology to myeloma proteins during the 1960s provided fundamental information about the structural nature of antibody diversity. Multiple myeloma patients produced large amounts of a single gamma globulin called myeloma protein that appeared to lack the heterogeneity of the molecules from normal individuals. They were each homogeneous and each differed from the others in the net charge. Myeloma proteins were invaluable in the resolution of antibody structure because they are produced in large quantities and are usually homogeneous in a patient, whereas in a normal immune response, the antibodies are heterogeneous in composition.

Moreover, some of these patients excreted a smaller protein (Bence Jones protein, first described in 1847 by Henry Bence-Jones) in their urine that was homogeneous and had unusual solubility properties during heat denaturation. When heated, the urine becomes cloudy in much the same fashion as urine containing albumen. However, on continued heating the urine becomes clear. Bence-Jones proteins were discovered to be monoclonal light chains quite independently of Edelman, and variable and constant regions, first seen in these Bence-Jones proteins, were crucial for a revolutionary change in the paradigm "One gene—one polypeptide"—again not by Edelman.

As Edelman (2004b) wrote: "Given my hypothesis about myelomas, the thought arose that perhaps Bence Jones protein was one of the chains of the myeloma protein that spilled into the urine because of its relatively low molecular weight (about 22,000)." Edelman decided to study these proteins, and after he compared reduced myeloma proteins from a number of different patients, he demonstrated that each protein, when reduced and alkylated and subjected to starch gel electrophoresis, had a unique migration pattern (Edelman and Poulik, 1961). Bence Jones proteins were simple excreted light chains. Edelman heated a sample of light chain obtained from normal human serum gamma globulins and demonstrated that they had the behavior of Bence Jones proteins becoming insoluble and then resolubilizing with continued heating (Edelman and Gally, 1962). In 1967, Putman et al. demonstrated that different Bence Jones proteins were different in their peptide sequence.

As Edelman (2004b) wrote: "The key experiment was to use the recently discovered starch gel electrophoresis to take a whole bunch of patients' urine samples, separate their Bence Jones proteins, and simultaneously separate and compare their serum proteins. When we ran the electrophoresis we demonstrated that the light chains were the Bence Jones proteins and they were pure and no two had the same mobility pattern." Edelman showed that each antibody chain has a tandem series of repeating homology units roughly 110 amino acid residues in length called immunoglobulin domains, which fold independently into a compact globular structure. All proteins that exhibit this structural motif belong to the immunoglobulin gene super family.

Since the discovery of Edelman, it has been clear and progressively established that antibodies are multifaceted proteins, capable of an extraordinary array of important functions. Not only major players in protection against invading pathogens, they play critical roles in inflammatory and autoimmune diseases, their induction is central to many vaccine strategies, and they are an unrivaled platform for engineering highly effective diagnostic and therapeutic reagents. Novel approaches to engineer effective therapeutic antibodies are also a key theme, including strategies to control antibody half-life and biodistribution.

REFERENCES

Burnet, F.M., 1959. The Clonal Selection Theory of Acquired Immunity. Vanderbilt University Press, Nashville.

Edelman, G.M., 1959. Dissociation of γ-globulin. J. Am. Chem. Soc. 81, 31–55.

Edelman, G.M., 1970. The covalent structure of a human γG-immunoglobulin. XI. Functional implications. Biochemistry 9, 3197–3204.

Edelman, G.M., 2004a. Interview with Gerald M. Edelman. Part I. Bioessays 26, 204–213.

Edelman, G.M., 2004b. Biochemistry and the sciences of recognition. J. Biol. Chem. 279, 7361–7369.

Edelman, G.M., Gally, J.A., 1962. The nature of Bence Jones proteins. J. Exp. Med. 116, 207–227.

Edelman, G.M., Poulik, M.D., 1961. Studies on structural units of the globulins. J. Exp. Med. 113, 861–884.

Edelman, G.M., Cunningham, B.A., Gall, W.E., et al., 1969. The covalent structure on entire γG-immunoglobulin molecule. Proc. Natl. Acad. Sci. U.S.A. 63, 78–85.

Hilschmann, N., Craig, L.C., 1965. Amino acid sequence studies with Bence Jones proteins. Proc. Natl. Acad. Sci. U.S.A. 53, 1403–1409.

Hirs, C.H., Moore, S., Stein, W.H., 1960. The sequence of the amino acid residues in per formic acid-oxidized ribonuclease. J. Biol. Chem. 235, 633–647.

Jerne, N.K., 1955. The natural selection theory of antibody formation. Proc. Natl. Acad. Sci. U.S.A. 41, 849–857.

Pauling, L., 1940. Theory of structure and process of formation of antibodies. J. Am. Chem. Soc. 62, 2643–2657.

Porter, R.R., 1959. The hydrolysis of rabbit γ-globulin and antibodies with crystalline papain. Biochem. J. 73, 119–126.

Press, E.M., Hogg, N.M., 1969. Comparative study of two immunoglobulin G-Fd-fragments. Nature. 223, 808–810.

Stretten, A., 2002. The first sequence. Fred Sanger and insulin. Genetics 162, 527–532.

CHAPTER 6

From the Discovery of Monoclonal Antibodies to Their Therapeutic Application: An Historical Reappraisal

6.1 THE DISCOVERY

As Cooper (2015) remembered: "The revolutionary technique of fusing a malignant plasma cell with normal B cell to form a hybridoma that produces monoclonal antibodies was discovered during the fertile period of B cell study in the 1970s."

The development by the Argentinian biochemist César Milstein (1927–2002) and his German postdoctoral fellow Georges Köhler (1946–95) (Figs. 6.1 and 6.2) in Milstein's lab at the "Molecular Research Council" "Laboratory of Molecular Biology" in Cambridge of the hybridoma technology has disclosed a new era in basic and applied oncology. For this discovery, in 1984, Milstein and Köhler shared the Nobel Prize for Medicine or Physiology together with Niels K. Jerne (1911–94) (Fig. 6.3) for "theories concerning the specificity in development and control of the immune system and discovery of the principle for production of monoclonal antibodies."

Milstein was interested in the mechanism of antibody diversity, and he believed that by determining the chemical structure of different antibodies, the antibody diversity problem could be solved. Milstein's lab was experienced in studying myeloma in mice. Using a technique developed earlier by Terry Potter, they had established a culture of rapidly proliferating tumor cells that produced immunoglobulins or antibodies.

Milstein suggested that Köhler investigated the antigen specificity of myeloma P3 antibodies. Instead of fusing two myelomas, Köhler decided to fuse one with mouse spleen cells, antibody producing B-cells. Spleen cells from immunized donors were fused with myeloma cells bearing a selection marker, and the fused cells were then cultured in a selective medium until visible colonies grew, and their supernatants were then screened for antibody production. For the first time,

Milestones in Immunology. DOI: http://dx.doi.org/10.1016/B978-0-12-811313-4.00006-1

Figure 6.1 A portrait of César Milstein.

Figure 6.2 A portrait of Georges Köhler.

Figure 6.3 A portrait of Niels K. Jerne.

large quantities of a pure antibody, specific for a single agent determinant, could be produced.

The key to success was the development of a selective technique to recover only fused cells, employing a mutant myeloma cell line deficient the enzyme hypoxanthine phosphoribosyl transferase. Without this enzyme, the cells would die in a medium containing hypoxanthine, aminopterine, and thymidine, but the hybrid cells would survive and could be selected, as the normal antibody-forming cell component of the hybrid would contribute the enzyme required (Kohler and Milstein, 1975, 1976)

The technique proposed by Köhler and Milstein is founded on three key principles: (1) each B cell produces only one antibody; (2) the lymphocytes used for the fusion are derived from donors that were sensitized with specific immunogens; (3) B cells can be immortalized into immunoglobulin-secreting in vitro cell lines.

Köhler and Milstein explained their collaboration in these extremely essential terms: "We agree that both conception and execution of the work was the result of close collaboration between us, with the skilled technical assistance of Shirley Howe. We are further convinced that the combined effect which results from such close collaboration was of a synergistic nature, synergistic effects taken to mean, as with monoclonal antibodies, effects which result from the combined action of two but which cannot be produced by the two separately. We both have a most pleasant memory of an exciting period in which a word, a comment, or a passing remark made by one had a resonant effect on the other. We do not want such happy memories, which have sealed a close friendship, to be disturbed by superficial interpretations of our individual recollections. It was a collaborative work. It was a collaborative paper. We do not want to make further comments." (Tansey and Catterall, 1994).

Köhler and Milstein did not patent this method, allowing the use of the hybridoma technology to academics and pharmaceutical industry for generation of future potential therapy. Initially work used myeloma cells which retained the capacity to secrete their own immunoglobulin products. Later, such fusion was replaced by myeloma variants that express only one endogenous chain or that fail to express immunoglobulin so that the fused cells secreted primarily or exclusively antibody of the desired specificity.

The monoclonal antibody technology has improved diagnostic applications including epitope specific immunoblotting, immunofluorescence, and immunohistochemistry.

6.2 THERAPEUTIC APPLICATIONS

Monoclonal antibody-mediated therapy started with mouse monoclonal antibodies, mouse to mouse-human chimeras, and later to humanized monoclonal antibodies. Antibodies of all types (murine, chimeric, and human) have been approved by Food and Drug Administration (FDA) and by other international agencies for the treatment of several pathologies.

In the late 1980s, murine monoclonal antibodies were in clinical development. The first monoclonal antibody approved by FDA for human use was a murine anti-CD3 monoclonal antibody, muromonab (OKT3), used for the treatment of organ transplant rejection (Kung et al., 1979).

However, murine monoclonal antibodies were often associated with allergic reactions and the induction of antidrug antibodies. To overcome these side effects, chimerization was developed in 1984 (Morrison et al., 1984). Chimeric mouse-human antibodies were enabled by grafting the entire antigen-specific domain of a mouse antibody onto the constant domains of a human antibody using genetic engineering techniques (Morrison et al., 1984).

Rituximab (Rituxan), a mouse-human chimeric monoclonal antibody against the B-cell lineage marker CD20 (expressed on the surface of normal B cells and on more than 90 percent of B-cell neoplasms, and from the pre-B-cell stage through terminal differentiation to plasma cells), was the first monoclonal antibody to be approved in 1997 by FDA for treatment of malignancy.

In most B cell non-Hodgkin's lymphoma (including chronic lymphocytic leukemia, mantle cell lymphoma, Waldenström macroglobulinemia, and marginal lymphoma) patients, rituximab causes a profound depletion of circulating B cells (Maloney et al., 1994). Normal B cells counts can decrease to zero after the initial infusion and recovery begins after 6 months and is complete by 9–12 months. The mechanism of action of rituximab includes antibody-dependent cytotoxicity, complement-dependent cytotoxicity, and chemo and radiosensitization of tumor cells (Eisenbeis et al., 2003).

Since then, no less than 15 distinct monoclonal antibodies have been approved for the treatment of hematologic and solid tumors, and between 1984 and 1988, both chimerization and humanization of monoclonal antibodies were achieved. In 2004, Cetuximab (Erbitux), a chimeric IgG1 directed against Her-1, a member of epidermal growth factor receptor (EGFR) family, was approved for the treatment of colorectal carcinoma and evaluated in patients with head and neck tumors that overexpress EGFR (Galizia et al., 2007; Kawaguchi et al., 2007). In experimental systems, treatment with cetuximab inhibits tumor cell proliferation and blocks the production of angiogenic factors.

Further development was the generation of fully human monoclonal antibodies, which was allowed through the advent of in vitro phage display technology and the generation of different mouse strains expressing human variable domains. Single chain Fv antibody phage display technology can be used to create a library of antibody genes that has subsequently been used and modified to build large human antibody libraries for therapeutic monoclonal antibodies generation.

Phage display technology enables selection of antibody fragments (e.g., scFv/Fab) with high affinity, specificity, and effector functions against various targets. However, such selection process itself is largely dependent upon various molecular factors such as methods for construction of phage library, phage/phagemid vectors, helper phage, host cells, and biopanning processes.

Molecular techniques were used to eliminate those portions in the murine immunoglobulin chains that are not involved in the binding of antigen and to replace them with the corresponding human sequences. The first humanized monoclonal antibody used as a therapeutic agent was an anti-CD25 monoclonal antibody to suppress reaction after transplantation. However, even fully completely evade the control by the host immune system, because they are normally capable to evoke an anti-idiotype response. Immunostimulatory monoclonal antibodies have been also developed to target the cancer cell/immune system crosstalk and related signaling pathways.

In 1998, Trastuzumab (Herceptin), a humanized IGg1 binding to the extracellular domain of the human Her-2/neu receptor, a member of EGFR family, was approved for the treatment of metastatic Her-2-overexpressing breast cancer (Hudis, 2007). In breast cancer, overexpression of Her-2 is associated to poor prognosis (Sledge, 2004). Trastuzumab induces the expression of antiangiogenic, suppresses proangiogenic factors, and mediates antibody-dependent cytoxicity. In 2000, Gemtuzmab ozogamicin (Mylotarg), a humanized IgG4 anti CD-33 (a glycoprotein receptor expressed on most myeloid leukemic blasts and leukemic progenitor cells, on normal and myelo-monocytic hematopoietic progenitor cells) conjugated to calicheamycin, a cytotoxic antibiotic, was approved by FDA for the treatment of patients with acute myeloid leukemia (Pagano et al., 2007). Gemtuzmab ozogamicin is an example of a monoclonal antibody conjugated with a toxin (immunotoxin). In 2001, Alemtuzumab (Campath), a humanized IgG1 monoclonal antibody anti-CD52 (a cellular surface glycoprotein) expressed by both normal and malignant B and T lymphocytes, was approved by FDA for the treatment of drug-resistant chronic lymphocytic leukemia, where disease in the blood and bone marrow responded more frequently than did nodal disease (Alinari et al., 2007). Alemtuzumab can induce tumor cell death through antibody-dependent cytotoxicity and complement-dependent cytotoxicity. Campath has been used also to deplete cells from allogenic transplantation grafts in patients with hematologic malignancies (Naparstek et al., 1999).

The first antiangiogenic agent approved by FDA for clinical use was bevacizumab (Avastin), a humanized version (Presta et al., 1997) of a murine monoclonal antibody against VEGF165 (Kim et al., 1993). Bevacizumab binds and neutralizes all human VEGF isoforms and bioactive proteolytic fragments. The clinical trial that resulted in FDA approval of Avastin in February 2004 was a randomized double-blind phase III study in which Avastin was administered in combination with bolus IFL (irinotecan, 5 fluorouracil, leucovirin) chemotherapy as first line therapy for previous untreated metastatic colorectal cancer (Hurwitz et al., 2004). Bevacizumab have been used in combination with conventional chemotherapy and/or targeted anticancer agents in other malignancies, including acute myeloid leukemia, multiple myeloma, head and neck squamous cell carcinoma, breast cancer, melanoma, hepatocellular carcinoma, pancreatic cancer, ovarian carcinoma, and prostate cancer. In 2006, FDA approved ranibizumab (Lucentis), an anti-VEGF antibody, as a therapy for neovascular age-related macular degeneration.

Bispecific include a group of monoclonal antibodies that can have multiple binding domains within the same construct that allow for interaction with two target antigens (Muller and Kontermann, 2010). The first bispecific monoclonal antibody for clinical use was in 2009 Catumaxomab, a chimeric (mouse and rat) monoclonal antibody specific for CD3 and epithelial cell adhesion molecule, used for the treatment of malignant ascites (Linke et al., 2010).

Among the fully human monoclonal antibodies, ipilimumab, tremelimumab, and MDX-1106 should be mentioned. Ipilimumab is a fully human IgG1 monoclonal antibody targeting the cytotoxic T lymphocyte antigen-4 on the surface of helper T cells and inhibiting the development of peripheral immune tolerance. Ipilimumab is currently approved for the treatment of melanoma (Acharya and Jeter, 2013). Similar to ipilimumab, tremelimumab, a human IgG2 monoclonal antibody, has been tested in several phases I–II clinical studies to evaluate its pharmacological profile in metastatic melanoma and in metastatic renal cell carcinoma (Camacho et al., 2009; Rini et al., 2011). MDX-1106 is a fully human IgG4 that specifically targets programmed death 1, a transmembrane receptor that mediates immunosuppressive functions in T cells (Kline and Gajewski, 2010). MDX-1106 has been tested in phase I–II studies for the therapy of different solid tumors (Wang et al., 2009; Brahmer et al., 2010). In addition, human immunoglobulin transgenic mice have been used to produce fully human antibodies, by disabling the ability of mice to produce their own murine antibodies and replacing that function with human antibody genes (Bruggemann et al., 1991). The transgenes are formed by fragments of the variable regions in germ lines, facilitating the capability to recombine the human antibodies. The first fully human monoclonal antibody developed from one of these systems was Vectibix, a human IgG2 antibody discovered using Ab-genix Xeno Mouse technology in 2006. Transgenic technology represents a powerful method for generating fully human monoclonal antibodies against a wide variety of drug targets, whereas recombinant technology continues to evolve, improving the pharmacodynamic and pharmacokinetic properties of antibody therapeutics, with the production of different antibody constructs or formats, such as bispecific antibodies, diabodies, and others, and different functional activities, such as catalysis, cellular internalization, and antigen-mimicking.

Autoimmune diseases, including rheumatoid arthritis, inflammatory bowel diseases, multiple sclerosis, and lupus erythematous, are another group of human diseases in which monoclonal are widely used. The most widely used are not only anticytokine antibodies, in particular antitumor necrosis factor alpha (adalimumab), but also anti-CD20 and anti-CD25.

6.3 CONCLUDING REMARKS

Before 1975, the capacity of antibodies to specifically target immunogens for therapeutic applications was not exploited due to the fact that antibody preparations were derived from polyclonal antisera. Monoclonal antibody-mediated therapy covers the fields of cancer, infectious diseases, transplantation, allergy, asthma, and some autoimmune diseases, and there are more than 250 therapeutic monoclonal antibodies undergoing clinical trials. Problems associated with immunogenicity, poor effector functions, and pharmacokinetics have been solved by chimerization and humanization. Five monoclonal antibodies, rituximab, infliximab, bevacizumab, trastzumab, and adalimumab, generated sales of over dollar 4 billion each in 2008.

Monoclonal antibodies are not an important modality for cancer treatment and have demonstrated antitumor activity in a broad spectrum of malignancies. Their toxicities are in general self-limited, including fever, nausea, fatigue, blood pressure fluctuations, and bronchospasm.

REFERENCES

Acharya, U.H., Jeter, J.M., 2013. Use of ipelimumab in the treatment of melanoma. Clin. Pharmacol. 5 (Suppl. 1), 21–27.

Alinari, L., Lapalombella, R., Andritsos, L., et al., 2007. Alemtuzumab (Campath 1H) in the treatment of chronic lymphocytic leukemia. Oncogene 26, 3644–3653.

Brahmer, J.R., Drake, C.G., Wollner, I., et al., 2010. Phase I study of single-agent anti-programmed death (MDX-1106) in refractory solid tumors: safety, clinical activity, pharmacodynamics and immunologic correlates. J. Clin. Oncol. 28, 3167–3175.

Bruggemann, M., Spicer, C., Buluwela, L., et al., 1991. Human antibody production in transgenic mice: expression from 100 kb of the human IgH locus. Eur. J. Immunol. 21, 1323–1326.

Camacho, L.H., Antonia, S., Sosman, J., et al., 2009. Phase I/II trial of tremelimumab in patients with metastatic melanoma. J. Clin. Oncol. 27, 1075–1081.

Cooper, M.D., 2015. The early history of B cells. Nat. Rev. Immunol. 15, 191–197.

Eisenbeis, C.F., Caligiuri, M.A., Byrd, J.C., 2003. Rituximab: converging mechanisms of action in non Hodgkin's lymphoma? Clin. Cancer. Res. 9, 5810–5812.

Galizia, G., Lieto, E., De Vita, F., et al., 2007. Cetuximab, a chimeric human mouse anti-epidermal growth factor receptor monoclonal antibody, in the treatment of human colorectal cancer. Oncogene. 26, 3654–3660.

Hudis, C.A., 2007. Trastuzumab—mechanism of action and use in clinical practice. N. Engl. J. Med. 357, 39–51.

Hurwitz, H., Fehrenbacher, L., Novotny, W., et al., 2004. Bevacizumab plus irinotecan, fluorouracil, and leucovorin for metastatic colorectal cancer. N. Engl. J. Med. 350, 2335–2342.

Kawaguchi, Y., Kono, K., Mimura, K., et al., 2007. Cetuximab induce antibody dependent cellular cytotoxicity against EGFR-expressing esophageal squamous cell carcinoma. Int. J. Cancer 120, 781–787.

Kim, K.J., Li, B., Winer, J., et al., 1993. Inhibition of vascular endothelial growth factor-induced angiogenesis suppresses tumour growth in vivo. Nature. 362, 841–844.

Kline, J., Gajewski, T.F., 2010. Clinical development of mAbs to block the PD1 pathway as an immunotherapy or cancer. Curr. Opin. Invest. Drugs 11, 1354–1359.

Kohler, G., Milstein, C., 1975. Continuous cultures of fused cells secreting antibody of predefined specificity. Nature. 256, 495–497.

Kohler, G., Milstein, C., 1976. Derivation of specific antibody-producing tissue culture and tumor lines by cell fusion. Eur. J. Immunol. 6, 511–519.

Kung, P., Goldstein, G., Reinherz, E.L., et al., 1979. Monoclonal antibodies defining distinctive human T cell surface antigens. Science 206, 347–349.

Linke, R., Klein, A., Seimetz, D., 2010. Catumaxomab: clinical development and future directions. MAbs 2, 129–136.

Maloney, D.G., Liles, T.M., Czerwinski, D.K., et al., 1994. Phase I clinical trial using escalating single-dose infusion of chimeric anti-CD20 monoclonal antibody (IDEC-C2B8) in patients with recurrent B-cell lymphoma. Blood. 84, 2457–2466.

Morrison, S.L., Johnson, M.J., Herzenberg, L.A., et al., 1984. Chimeric human antibody molecules: mouse antigen-binding domains with human constant region domains. Proc. Natl. Acad. Sci. U.S.A. 81, 6851–6855.

Muller, D., Kontermann, R.E., 2010. Bispecific antibodies for cancer immunotherapy: current perspectives. BioDrugs 24, 89–98.

Naparstek, E., Delukina, M., Or, R., et al., 1999. Engraftment of marrow allografts treated with Campath-1 monoclonal antibodies. Exp. Hematol. 27, 1210–1218.

Pagano, L., Fianchi, L., Caira, M., et al., 2007. The role of Gemtuzumab Ozogamicin in the treatment of acute myeloid leukemia patients. Oncogene 26, 3679–3690.

Presta, L.G., Chen, H., O'Connor, S.J., et al., 1997. Humanization of an anti-vascular endothelial growth factor monoclonal antibody for the therapy of solid tumors and other disorders. Cancer Res. 57, 4593–4599.

Rini, B.I., Stein, M., Shannon, P., et al., 2011. Phase dose-escalation trial of tremelimumab plus sunitinib in patients with metastatic renal cell carcinoma. Cancer 117, 758–767.

Sledge Jr., G.W., 2004. HERe-2 stay: the continuing importance of translational research in breast cancer. J. Nat. Cancer Inst. 96, 725–727.

Tansey, E.M., Catterall, P.P., 1994. Monoclonal antibodies: a witness seminar in contemporary medical history. Med. Hist. 38, 322–327.

Wang, W., Lau, R., Yu, D., et al., 2009. PD1 blockade revers the suppression of melanoma antigen-specific CTL by $CD4^+$ CD25(Hi) regulatory T cells. Int. Immunol. 21, 1065–1077.

CHAPTER 7

Sir Frank Macfarlane Burnet and the Clonal Selection Theory of Antibody Formation

7.1 INTRODUCTION

Many accounts have been previously published concerning the ontogeny of clonal selection theory of antibody formation. Lederberg (1988) published his reflections on Charles Darwin and Paul Ehrlich entitled "Ontogeny of the clonal selection of antibody formation." In 1995, The "FASEB Journal" published a special article of Forsdyke (1995) entitled "The origins of the clonal selection theory of immunity as a case study for evaluation in science." In 2002, "Nature Immunology" published a commentary of Silverstein (2002) entitled "The clonal selection theory: what it really is and why modern challenges are misplaced." In 2007, in occasion of the 50th anniversary of the publication of Frank Macfarlane Burnet's clonal selection theory, "Nature Immunology" published an Editorial, an Essay signed by Gustav J.V. Nossal, an Historical commentary containing the reproduction of the original version of the 1957 Burnet's paper (Editorial, 2007; Nossal, 2007; Hodgkin et al., 2007). In the same year "Nature Reviews in Immunology" published a Viewpoint entitled "Reflections on the clonal-selection theory" (Cohn et al., 2007). Finally, Gordon Ada (2008) published a Landmark entitled "The enunciation and impact of Macfarlane Burnet's clonal selection theory of acquired immunity."

7.2 BIOGRAPHICAL NOTE

Sir Frank Macfarlane Burnet was born in Traralgon, in eastern Victoria, Australia, on September 3, 1899. He was educated at the Victoria State Schools and at Geelong College, completing his medical course at the University of Melbourne, where he graduated Medical Doctor in 1923. In 1926, he was awarded a Beit Fellowship for Medical Research and worked for a year at the Lister Institute, London. In 1932, he spent a year at the National Institute for Medical Research, Hampstead, London. Otherwise, he has worked

Milestones in Immunology. DOI: http://dx.doi.org/10.1016/B978-0-12-811313-4.00007-3

continuously at the Hall Institute in Melbourne and from 1944 to 1965 he was Director of this Institute and Professor of Experimental Medicine in the University of Melbourne (Fig. 7.1).

From 1951 to 1956, Burnet concentrated on studies of the genetics of influenza virus. In parallel with his work on virology, Burnet et al. (1941) had always been interested in the immune response, and in 1941, he published a monograph analyzing the nature of antibody formation and enunciated in this book the hypothesis of acquired immunological tolerance. In 1948, he proposed a new hypothesis on antibody production based on analogies with adaptive enzymes (Burnet and Fenner, 1948). In 1957, Burnet changed the direction of his own work and that of the Institute, abandoning virology and concentrating on immunology instrumental in attracting some of the great names in immunology to the institute.

In 1960, Burnet received the award of the Nobel Prize in Physiology or Medicine jointly with Sir Peter Medawar, for the discovery of the acquired immunological tolerance. In 1965, Burnet retired from directorship of the Institute, and Doctor Gustav Nossall was

Figure 7.1 A portrait of Sir Frank Macfarlane Burnet.

appointed as Director. To mark the occasion, the "Ciba Foundation" organized a symposium on the thymus in Melbourne.

After retirement, he moved into the School of Microbiology in the University of Melbourne. During the 12 years he was at the University of Melbourne, Burnet produced 13 books, initially on immunology and subsequently on human biology, aging and cancer, as well as a fourth edition of his first book.

In 1969, and again in 1974, international symposia were organized by Nossal to celebrate his 70th and 75th birthdays. In 1978, at the age of 78, Burnet left the School of Microbiology and moved to his home, where he produced two more books. In November 1984, he was operated for rectal cancer, but secondary lesion was discovered early in August 1985, and he died on August 31, 1985.

7.3 THE INSTRUCTIVE THEORIES OF ANTIBODY PRODUCTION

During the 1930s, Breinl and Haurowitz (1930) and Mudd (1932) proposed a hypothesis to account for antibody production which was clarified and reformulated by Pauling (1940). The "instructive" theory, as it was proposed by Linus Pauling, arose from the universally accepted concept that the antibody repertoire had to be transcendental to protect the animal. As the number of genes had to be lower than the number of antigens that are recognized, the only solution seemed to be the "instructive" theory. According to the "instructive" theory, the antibody was synthesized, or according to Pauling folded, in specific ways in spatial contact with the antigenically significant parts of the antigen, which acted as a template. No supporting findings were found for this concept, but much contrary evidence accumulated.

Burnet et al. (1941) summarized his views on antibody production in a monograph. Due to the apparently almost infinite variety of antibodies, Burnet accepted an instructive hypothesis but suggested that the antigen impressed a complementary pattern not on the globulin molecule but on some cellular components. In the introduction, Burnet identified the problem of antibody synthesis as linked to key biological questions: the conditions governing protein synthesis in the living cell and the capacity of biological systems to be modified in reaction. This second point, the capacity of an organism to remember its first encounter with a given antigen and to elaborate an accelerated secondary immune response, had brought Burnet to reflect on immunological behavior of young animals. Burnet's interpretation of the finding that chick embryo was able to tolerate foreign tissues was: "in some ways the embryonic cells seem to be unable to recognize and resent contact with foreign material in the way adults cells do."

Burnet and Fenner (1948) published a short review in which they ascribed antibody production to the "existence and modification of self-replicating units within the antibody producing cells" and included "the antibody-producing mechanism into line with the growing number of cytoplasmic entities which have demonstrably, or probably, power of self-replication in response to environmental rather than nuclear stimuli."

As Talmage (1986) remembered: "In 1949, Burnet and Fenner published the first edition of a small book called 'Production of antibodies.' In it, they strongly attacked the current chemical approach to immunology and the isolation of immunology from the mainstream of biology. They looked to adaptive processes in bacteria for a model of antibody formation and introduced the term 'protein synthesizing unit' (ribosomes were unknown at that time). They also attempted to explain Owen's red cell chimeras with a self-marker theory of immunological tolerance. Thus, the little book played a large role in changing the conceptual framework through which antibody formation was viewed."

In a second edition of the monograph (Burnet and Fenner, 1949), Burnet developed a new hypothesis for the process of antibody production based on an analogy with adaptive enzyme process in bacteria, in which bacteria produce large amounts of an enzyme that is capable of breaking down to specific sugar that is placed in culture medium. Burnet suggested that an antigen might induce modifications of those enzymes involved in globulin synthesis so that a protein with the required specificity might be formed. This hypothesis concerned the manner in which the body normally failed to make antibodies to its own components, the "self-marker" concept, according to which the distinction between the "self" and the "not-self" is based on a limited number of recognizable components in each cell, the combination of which constitute the specific "self-pattern" of the organism. Burnet reported data concerning the effect that mice and calves exposed continuously to antigens during embryonic life failed to produce antibodies if exposed to these antigens in adult life. He concluded: "If in embryonic life expandable cells from a genetically distinct race are implanted and established, no antibody response should develop against the foreign cell antigen when the animal takes on independent existence."

In a book published in 1953, Burnet (1953) proposed that antigens induced the formation of stable cytoplasmic synthetic units, "almost a case of inheritance of acquired characteristics." In 1956, recognizing the growing importance of nucleic acids in protein synthesis, Burnet (1956) proposed a complicated model in which antigens associated with either cytoplasmic RNA or with DNA served as a template for synthesis of a new protein.

7.4 THE SELECTIVE THEORIES OF ANTIBODY PRODUCTION

Ehrlich (1900) published a selective theory of antibody formation, called the "side chain theory." The theory proposed that the antibody located on cell surface could serve as a receptor for antigen. Following reaction with a foreign antigen, the receptor/antigen complex would be discarded from the cell surface. The affected cell overproduces more side chains which would become circulating antibodies. Ehrlich (1900) wrote: "In the course of immunization the cells become, so to say, educated or trained to reproduce the necessary side-chains in ever, increasing quantity." Ehrlich differed from his contemporary Elie Metchinkoff who ascribed the production of antibodies to macrophages. Ehrlich suggested that this function might be a specialized characteristic of "hematopoietic tissue."

The key feature of Ehrlich's model was that there was a preexisting repertoire of specificities for a variety of antigens. The antigens would act to select from among the specificities. He understood that antibody molecules have a distinct structure, and that parts of the molecule that react with complement might differ from parts reacting with specific antigens. He also recognized that antibodies themselves are potential antigens and that distinct anti-antibodies might be raised against different parts of the antibody molecule and introduced the idea of a mechanism of self/not-self-discrimination ("horror autotoxicus").

Ehrlich's (1900) main problem was: "It would not be reasonable to suppose that were present in the organism many hundreds of atomic groups destined to unite with toxins, when the latter appeared, but in function really playing no part in the process of normal life, and only arbitrarily brought into relationship with them by the will of the investigator. It would indeed by high superfluous, for example, for all our native animals to possess in their tissues atomic groups deliberately adapted to unit with abrin, ricin, and crotin, substances coming from the far distant tropics."

As Talmage (1986) pointed out: "The side-chain theory should be have launched the beginnings of cellular immunology. Instead was rapidly forgotten. Like Mendel, Ehrlich was ahead of his time. To the scientists of his day, the side-chain theory seemed so ridiculous that it was not worth considering. When Landsteiner showed that antibodies could be made to newly synthesized chemicals, the side-theory and the study of cells were dropped from consideration of immunologists for more than 30 years."

Breinl and Haurowitz (1930) proposed the first instructional theory, referred as the direct template model: "We assume that antigens interact in an early stage in the process of globulin formation, at the stage when non specific components are combined into a specific globulin molecule under the direction of the antigen. Instead of normal globulin, a globulin of a special structure – the antibody – will be generated .... The formation of antibodies would therefore not be an unusual occurrence, but would be nothing more than the formation of normal globulin under special circumstances that are conditioned by antigens.

The long time interest of Burnet in the theory of antibody production had been stimulated by a paper published by Jerne (1955) that proposed a "natural selective theory" for the process, rather than the "instructive" theory. In his paper entitled "The natural selection theory of antibody formation", Jerne (1955) wrote: "Among the comparatively small number, perhaps a few thousand, of antigen–antibody systems investigated, cross-reactions are by no means rare, suggesting that the number of specific configurations which globulin molecule can exhibit is large but limited. Since normal mammalian serum contains more than 10^{17} globulin molecules per millimeter, these may include a million 10^{11} fractions of different specificity. This would seem an amply sufficient number."

According to Jerne, the function of an antigen was to combine with those globulins with which it made a chance fit and to transport the selected globulins to antibody-producing cells, which would then make mainly identical copies of the globulin presented to them. Jerne proposed that antigen–antibody complexes are taken into cells where the antibody is then replicated. Jerne's concept that antibodies were natural globulins was attractive, but his concepts of globulin randomization and replication, like nucleic acids, were incompatible with new information that was available on DNA. Jerne's (1955) "natural selective theory" was basically a revision of Ehrlich's "side chain theory." Both theories held that antibodies, not cells, were selected by antigens.

Jerne (1955) proposed a potential mechanism for the induction of tolerance: "If this small spontaneous production of globulin took place mainly in embryonic and early independent life ... the early removal of a specific fraction of molecule might lead to the permanent disappearance of this type of specificity .... The absence from the circulation of such antibodies would, in turn, prevent response to a alter antigenic stimulus of this type." In the opinion of Talmage (1986): "Jerne's theory had several weakness. In addition to its inability to explain the cellular basis of immunologic memory, his postulation of protein replication was without precedent and at odds with the developing consensus that the ability to replicate was the exclusive property of nucleic acids. Nevertheless, it was a simple matter to substitute randomly diversified cells for Jerne's randomly diversified globulin molecules and thus to develop a cell selection theory of antibody formation."

Burnet rejected Jerne's theory based on the implausibility of a self-replicating antibody molecule and instead suggested that the B-cell receptor was located on cells that replicate.

David Talmage, replacing randomly diversified globulins with randomly diversified cells, provided a "cell selection theory" of antibody formation, and his working hypothesis favored a selective model and proposed that the unit responsible for expansion was the antibody-producing cell itself. In a review of 1957, he wrote. "... it is tempting to consider that one of the multiplying units in the antibody response is the cell itself. According to this hypothesis, only those cells are selected for multiplication whose synthesized product has affinity for the antigen injected. This would have the disadvantage of requiring a different species of cell for each species of protein produced, but would not increase the total amount of configurational information required on the hereditary process. (...) The cellular hypothesis is compatible with current concepts that the configuration of a protein molecule is determined solely by information contained in the hereditary units of the cell, the nucleic acid." (Talmage, 1957).

According to Talmage, to have selection there must be diversity so the concept was based on the idea that there were many different cells in the body that made antibodies, each making a different molecule or species of antibody. So, the clonal selection theory is basically a combination of the selection hypothesis and the cell hypothesis. Talmage discussed supporting evidence from the kinetics of antibody response, from immunological memory, and from the fact that myeloma tumors often result in a massive production of one globulin randomly selected from the family of normal globulins. Talmage never received the recognition he deserved for his seminal contribution.

7.5 THE CLONAL SELECTION THEORY

Following on Jerne's and Talmage's ideas, Burnet suggested the clonal selection approach. The basis of clonal selection theory is that the specific capacity of a cell to react immunologically, either as a cell or a producer of antibody, is conferred on by genetic processes and not by the intrusion of a pattern from the antigen. The clonal selection theory advanced the concept that antibodies were natural globulins that possessed an affinity for antigens and are selected from a large group of preexisting globulins.

Burnet published a short paper dated October 21, 1957, published in the "Australian Journal of Science", which was little more than a newsletter published by the "Australian and New Zealand Association for the Advancement of Science", and described as a "preliminary account", where he cited Talmage's paper: "Talmage has suggested that Jerne's view is basically an extension of Ehrlich's side chain theory of antibody production and that replicating elements essential to any such theory were cellular in character *ab initio* rather than circulating protein which can replicate only when taken into an appropriate cell. Talmage does not elaborate this point of view but clearly accepts it as the best basis for the future development of antibody theory. Before receiving Talmage's review we had adopted virtually the same approach but had developed it from what might be called a 'clonal' point of view." (Burnet, 1957).

As Nossal (2007) pointed out: "Burnet speculated that the small lymphocyte population of the body is actually a repertoire of specificities, with each bearing on its surface just one kind of natural antibody. During immunization, antigen selects cells with the corresponding specificity for multiplication and differentiation into antibody secreting status. This accounts for an exponential rise in antibody concentrations and an increased number of the 'right' cells to respond to a second immunization."

Burnet assumed that in a population of mesenchymal cells that are a certain number of different clones and that it is characteristic of any such clone that is composed of cells which are immunologically competent toward a certain antigenic determinant. If the antigenic determinant makes effective contact with such a cell, it is stimulated to proliferate (clonally expand) and under appropriate circumstances some of the descendant cells will be converted into plasma cells and produce antibodies, that combine with the antigen. If the antigen is part of the surface of a virus or bacterium, than the antibody labels that organism as foreign ("not self"). The organism is then ingested by phagocytic cells and degraded. In this "preliminary account" but in the meantime a "fundamental paper", Burnet wrote: "Among [antibodies] are molecules that can correspond probably with varying degrees of precision to all, or virtually all, the antigenic determinants that occur in biological material other than characteristic of the body itself. Each type of pattern is a specific product of a clone [lymphocytes] and it is the essence of a hypothesis that each cell automatically has available on its surface representative reactive sites equivalent to those of the globulin they produce. (...) It is assumed that when an antigen enters the blood or tissue fluids it will attach to the surface of any lymphocyte carrying reactive sites which correspond to one of its antigenic determinants. (...) It is postulated that when antigen–[antibody] contact takes place on the surface of a lymphocyte the cell is activated to settle in an appropriate tissue. (...) and there undergo proliferation to produce a variety of descendents. In this way, preferential proliferation will be initiated of all those clones whose reactive sites correspond to the antigenic determinants on the antigen used. The descendents will [be] capable of active liberation of soluble antibody and lymphocytes which can fulfill the same functions as the parental forms." (Burnet, 1957).

In the book published in 1959, Burnet clarified several points in the theory, and expanded upon the earlier hint on the importance of somatic mutation. He developed further the subsidiary hypothesis of clonal abortion during fetal life to explain tolerance to self-antigens (Burnet, 1959). He described in detail how the clonal selection theory could explain a broad range of immunological phenomena, including immunological memory, original antigenic sin, the effects of adjuvants, mucosal immunity, natural antibodies, and autoimmunity.

7.6 EVIDENCE SUPPORTING THE THEORY

Initially, there was opposition to the idea that the body could make enough different natural globulins to react specifically with every conceivable antigen. In response to this opposition, Talmage (1959) demonstrated how an almost unlimited number of different combinations of approximately 50,000 different globulins might explain immunological specificity.

Gordon Ada and Nossal collaborate to device a microscopic assay to test Burnet's concept. It was based on the ability of flagella-specific antibody, on addition to bacteria cultures, to prevent bacterial migration as observed under the microscope. Individual rats were immunized with two serologically distinct *Salmonella* flagella to establish if single plasma cells from the immunized rats produced antibodies to both flagella (double producers) or only to one. Nossal and Lederberg (1958) published a paper in 1958 with results from 62 positive cells showing that no double producers had yet been found and that the single antibody-producing cells in culture made only one antibody. Nossal (1960) confirmed these data in his PhD thesis, reporting that following examination of nearly 1500 anti-flagella-producing cells, no double producers were found. White (1958) found no double producers in his system, and Attardi et al. (1958) using a method based on the inactivation of bacterial phages, demonstrated that about 10% of active cells were double producers. A new technology introduced by Jerne, the plaque assay (Jerne and Nordin, 1963) allowed several groups to show that individual cells from donors who were immunized with multiple antigens made antibodies specific for only one of those antigens. Colonies of antibody-forming cells could be found in the spleen of animals given whole body gamma radiation and injected with small numbers of spleen cells (Palyfair et al., 1965). Raff et al. (1973) showed that incubation of lymphocytes with antigen could aggregate all the surface immunoglobulin on antigen-binding cells and this indicated that only immunoglobulin on the surface of these cells was antibody of a single specificity.

In the 1970s, cellular immunology itself exploded and molecular immunology also advanced rapidly. The structure of immunoglobulin and its genetic basis were completely developed. In 1976, following the development of the hybridoma technology it was possible to develop a technique to make large amounts of a monoclonal antibody (Kohler and Milstein, 1975). In this process, the antibody-secreting cells, which have a short half-life, are fused with myeloma cells, resulting in an immortal line of antibody-secreting cells. Jerne, George J.F. Kohler, and Cesar Milstein shared the 1984 Nobel Prize in Physiology or Medicine for this achievement.

7.7 CONCLUDING REMARKS

At a symposium on antibodies held at Cold Spring Harbor in 1967, Burnet delivered the opening address with a paper outlining the historical background to the evolution of his clonal selection theory. He wrote: "Jerne postulated that gamma globulin molecules are continuously being synthesized in an enormous variety of different configurations. The origin of the diversity is left unexplained. When an antigen intrudes into the body, sooner or later globulin molecules of the appropriate natural pattern will become attached to the antigenic molecules or particles. The complex is then taken up by phagocytic cell where, by hypothesis, the globulin can be released from the antigen. Such globulin molecules either in the macrophage of after transfer to another cell were said to serve as a 'signal for the synthesis or reproduction of molecules identical to those introduced, i.e. of specific antibodies.' Even in 1955 this seemed wholly inadmissible. Most other aspects of the new theory were highly acceptable but the basic flaw seemed to be a fatal one." (Burnet, 1967).

In the same symposium, Jerne (1967) acknowledged how Burnet's scientific originality had contributed to the modern immunology: "Sir Macfarlane Burnet must have been pleased not only to witness at this symposium the vindication of his clonal selection theory of acquired immunity, but also to see how his stimulating ideas have led to a great proliferation of immunologists and to know that the fate of immunology is deposited in so many capable hands."

In a paper published in 1988 and dedicated to the memory of Burnet, Lederberg (1988) wrote: "By the 1967 Cold Spring Harbor Symposium, the clonal selection theory was an undeniable fundament for almost every investigation of the chemistry of antibodies or the biology of immunocytes. It was also clear that further progress would depend on the propagation of antibody-forming cells as clones."

It was primarily for the formulation of the clonal selection theory that Burnet was included in the London *Sunday Times* 1969 list of the 1000 people who had made the twentieth century. His lasting contribution was expressed in this form: "By turning immunological theory upside down he has kept the world's immunologists busy and happy for the last decade."

In 1967, Jerne explained that there were two kinds of immunologists, who hardly spoke to each other: the "trans-immunologists," who were interested in the structure of the antibody molecule and its binding to the antigen, and the "cis-immunologists," who were interested in the events following antigen exposure. The consequence of acceptance of the clonal theory of antibody synthesis on the other hand, and the progress in understanding the mechanism of protein synthesis and the role of cellular receptors in the regulation of immune phenomena on the other, reduced the distance between the cis- and trans-immunologists.

REFERENCES

Ada, G., 2008. The enunciation and impact of Macfarlane Burnet's clonal selection theory of acquired immunity. Immunol. Cell. Biol. 86, 116–118.

Attardi, G., Cohn, M., Horibata, K., et al., 1958. On the analysis of antibody synthesis at cellular level. Bacteriol. Rev. 23, 213–223.

Breinl, F., Haurowitz, F., 1930. Chemische Untersuchung des Prazipitates aus Hamoglobin and Anti-Hamoglobin-Serum and Bemerkungen ber die Natur der Antikorper. Z. Phyisiol. Chem. 192, 45–55.

Burnet, F.M., 1953. Natural History of Infectious Diseases. Cambridge University Press, Cambridge.

Burnet, F.M., 1956. Enzyme, Antigen and Virus. Cambridge University Press, Cambridge.

Burnet, F.M., 1957. A modification of Jerne's theory of antibody production using the concept of clonal selection. Aust. J. Sci. 20, 67–69.

Burnet, F.M., 1959. The clonal selection theory of acquired immunity. Cambridge University Press, Cambridge.

Burnet, F.M., 1967. The impact on ideas of immunology. Cold Spring Harb. Symp. Quant. Biol. 32, 1–8.

Burnet, F.M., Fenner, F., 1948. Genetics and immunology. Heredity. 2, 289–324.

Burnet, F.M., Fenner, F., 1949. The production of antibodies, Monograph of the Walter and Eliza Hall Institute, Melbourne, second ed. Macmillan, Melbourne.

Burnet, F.M., Freeman, M., Jackson, A.V., et al., 1941. The production of antibodies: a review and theoretical discussion. Monograph from the Walter and Eliza Hall Institute of Research in Pathology and Medicine, No 1. Macmillan, Melbourne.

Cohn, M., Mitchison, N.A., Paul, W.E., et al., 2007. Reflections on the clonal-selection theory. Nat. Immunol. 7, 823–830.

Editorial, 2007. Sir Frank Macfarlane Burnet, 1899-985. Nat. Immunol. 8, 1009.

Ehrlich, P., 1900. On the immunity with special reference to cell life. Proc. R. Soc. London 66, 424–448.

Forsdyke, D.R., 1995. The origins of the clonal selection theory of immunity as a case study for evaluation in science. FASEB J. 9, 164–166.

Hodgkin, P.D., Heath, W.R., Baxter, A.G., 2007. The clonal selection theory: 50 years since the revolution. Nat. Immunol. 8, 1019–1023.

Jerne, N.K., 1955. The natural selection theory of antibody formation. Proc. Natl. Acad. Sci. U.S.A. 41, 849–857.

Jerne, N.K., 1967. Waiting for the end: a summary. Cold Spring Harb. Symp. Quant. Biol. 32, 591.

Jerne, N.K., Nordin, A.A., 1963. Plaque formation in agar by single antibody-producing cells. Science 140, 405.

Kohler, G., Milstein, C., 1975. Derivation of specific antibody-producing tissue culture and tumor lines by cell fusion. Eur. J. Immunol. 6, 511–519.

Lederberg, J., 1988. Ontogeny of the clonal selection theory of antibody formation. Reflections on Darwin and Ehrlich. Ann. N. Y. Acad. Sci. U.S.A. 546, 175–187.

Mudd, S., 1932. A hypothetical mechanism of antibody formation. J. Immunol. 23, 423–434.

Nossal GJV. PhD Thesis. University of Sidney, 1960.

Nossal, G.J.V., 2007. One cell-one antibody: prelude and aftermath. Nat. Immunol. 8, 1015–1017.

Nossal, G.J.V., Lederberg, J., 1958. Antibody production by single cells. Nature 181, 1419–1420.

Palyfair, J.H.L., Papermaster, B.W., Cole, L.J., 1965. Focal antibody production by transferred spleen cells in irradiated mice. Science 149, 998–1000.

Pauling, L., 1940. A theory of the structure and process of formation of antibodies. J. Am. Chem. Soc. 62, 2643–2657.

Raff, M.C., Feldman, M., de Petris, S., 1973. Monospecificity of bone marrow derived lymphocytes. J. Exp. Med. 137, 1024–1030.

Silverstein, A.M., 2002. The clonal selection theory: what it really is and why modern challenges are misplaced. Nat. Immunol. 3, 793–796.

Talmage, D.H., 1959. Immunological specificity: unique combinations of selected natural globulins provide an alternative to the classical concept. Science 129, 1643–1648.

Talmage, D.H., 1986. The acceptance and rejection of immunological concepts. Ann. Rev. Immunol. 4, 1–11.

Talmage, D.W., 1957. Allergy and immunology. Ann. Rev. Med. 8, 239–256.

White, R.G., 1958. Antibody production by single cells. Nature 182, 1383–1384.

FURTHER READING

Wolstenholme, G.E.W., Porter, R., 1966. Ciba Foundation Symposium: The Thymus, Experimental and Clinical Studies. Churchill, London.

CHAPTER 8

Peter Brian Medawar and the Discovery of Acquired Immunological Tolerance

8.1 BIOGRAPHIC NOTE

Peter Brian Medawar was born on February 28, 1915, Rio de Janeiro, Brazil, to businessperson Nicholas Medawar and the former Edith Muriel Dowling. After the conclusion of the First World War in 1918, the family moved to England. Medawar studied zoology at Magdalen College, Oxford, completing his undergraduate studies with first class honors in 1935. In the same year, he accepted an appointment as Christopher Welch Scholar and Senior Demonstrator at Magdalen College. He was named a fellow at Magdalen in 1938 and remained at Oxford until 1947, when he accepted an appointment as Mason Professor of Zoology, at the University of Birmingham.

As Park (2010) pointed out: "During this period, he learned to investigate the living organism as a constantly changing entity, whose nature could be revealed primarily through careful observation and systematic analysis that often relied upon mathematical tools (...) A major influence on this academic stand point was the British biologist D'Arcy Thompson, who wrote 'On Growth and Form' (1917). Through his publication and personal correspondence, Thompson taught Medawar that the growth and structural relationship of various living organisms could be mathematically traced and analyzed."

When World War II broke out in Europe, the Medical Research Council asked Medawar to concentrate his research on tissue transplants, primarily skin grafts. In these years, Medawar together with Rupert Billingham and Leslie Brent at the University College in London investigated the use of skin grafts to determine whether cattle twins were monozygotic or dizygotic. He observed that the rejection time for donor grafts was noticeably longer for initial grafts, compared with those grafts that were transplanted for a second time. Medawar formed the opinion that the body's rejection of skin grafts was immunological in nature.

Milestones in Immunology. DOI: http://dx.doi.org/10.1016/B978-0-12-811313-4.00008-5

From 1951 to 1962, Medawar served as professor of Zoology and Comparative Anatomy at University College London. In 1962, he became the director of the National Institute for Medical Research, Mill Hill (Fig. 8.1), where he continued his study of transplants and immunology. In 1960, Medawar was awarded the Nobel Prize in Physiology or Medicine with Sir Frank Macfarlane Burnet (Fig. 8.2) for their discovery of acquired immunologic tolerance. Medawar's Nobel prize award was in recognition of the significance of his 1953 and 1956 papers (Billingham et al., 1953, 1956, the alphabetical order of their names was Medawar's convention and acknowledgment of their team work) on induction of transplantation tolerance, experimentally providing supporting evidence for Burnet's hypothesis of self/non-self-discrimination. In 1969, Medawar suffered a stroke, the first of several, that left him partially paralyzed and forced him to step down as director in 1972. On October 2, 1987, at the age of 72, Medawar died at the Royal Free Hospital in London as a consequence of a last severe stroke. He was made a fellow of the Royal Society in 1949 and received many other honors throughout his career. Medawar was also a philosopher and a gifted science communicator.

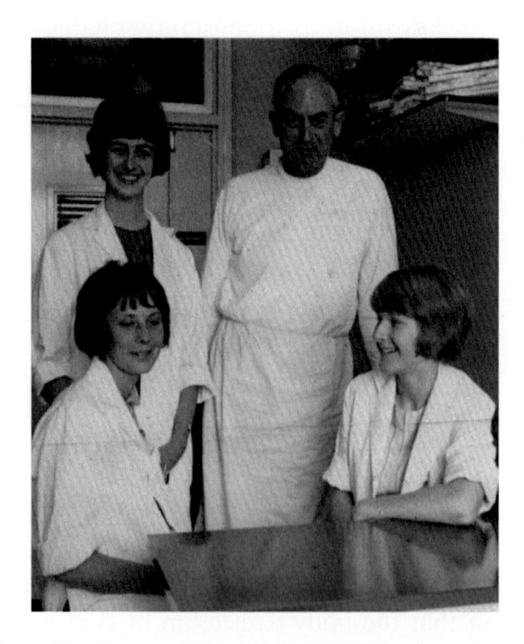

Figure 8.1 Peter Medawar with colleagues at National Institute for Medical Research.

Figure 8.2 A portrait of Sir Frank Macfarlane Burnet.

8.2 THE INFLUENCE OF BURNET

In 1945, Raymond Owen (Fig. 8.3) discovered that chimerism in cattle twins protected red blood cells (RBCs) of both animals in each of them. He made the observation that nonidentical (dizygotic) twin cattle, which shared the same placental circulation and whose circulations were thereby linked, grew up with appreciable numbers of RBCs from the other twin in their blood: if they had not shared the same circulation at birth, red cells from the twin injected in adult life would be rapidly eliminated by an immunological response (Owen, 1945). This paper was significant with regards the hemopoietic stem cell research as well. When Medawar was awarded the Nobel Prize, he wrote a letter to Owen in which he declared: "I think it is very wrong that you are not sharing in this prize; the only consolation is that all your professional colleagues have a perfectly clear understanding of the fact that you started it all. I have been tortured by doubts as to whether or not that is a fact that I myself have made clear enough in my own publications – so I looked up our big paper on tolerance in the Phyl. Trans. of 1956, and don't think we can reproach ourselves. The fact of the matter is that luck plays altogether too high a part in these awards – they ought at

Figure 8.3 A portrait of Raymond Owen.

least consult the intended recipient before the award is made, for he should know best where credit is due." (Brent, 1997).

From this finding, Burnet and his colleague Frank Fenner conceived the notion that potential antigens which reach the lymphoid cells during their developing immunologically immature phase in the perinatal period can specifically suppress any future response to that antigen when the animal reaches immunological maturity.

As Burnet (1971) said: "In the first edition of 'Biological Aspects of Infectious Disease' (1940), I first used the concept that once the simplest animals evolved, the related requirements for mutation (by the digestion of other organisms) and for protection from bacterial infection, required a capacity to distinguish between the chemical structure characteristic of self' and any sufficiently different chemical structure which is recognized as 'not self'. Self' must remain undamaged by the enzymes and other mechanisms that can digest and destroy 'not self'. This is the concept basic to modern immunology but no one applied it seriously until 1949. In that year, Fenner and I published a second edition of an Institute monograph on 'The production of antibodies.' In this there was the first clear recognition that the differentiation of self from not-self was very important in immunology and that, to a large extent, it was developed in birds and mammals during embryonic life."

In 1949, Burnet and Fenner published the second edition of their book entitled "The Production of Antibodies." The core of this book was the development of the concept of self/nonself-discrimination: "If in embryonic life expandable cells from a genetically distinct race are implanted and established, no antibody response should develop against the foreign cell antigen when the animal takes an independent existence." In other words, unresponsiveness to the body's own constituents ("self") could be established and thereby enable the lymphoid cells to make the distinction between "self" and "nonself."

In the Introduction to this monograph, Burnet identified the problem of antibody synthesis as linked with two key biological questions: the condition governing protein synthesis in the living cells, and the capacity of an organism to remember its first encounter with a given antigen and to mount an accelerated "secondary" immune response. Murphy (1913) had demonstrated that the chick embryo was able to tolerate foreign tissues. Murphy predicted the absence of an immune response in the chick chorioallantoic membrane at the time in which he performed the experiments. He wrote, "Apart for the thin continuation of the chick membrane which covers the tumor and the ingrowth of vessels with their scant accompanying stroma, there is no histological evidence of reactions on the part of the embryo to the invasion of foreign tissue" (Murphy, 1913). This observation was confirmed after the 1950s by a number of researches concerning the morpho-functional characterization of the immune system in the chick embryo. Early lymphoid cells deriving from the yolk sac and spleen are usually recognizable in the thymus on day 8 and in the bursa of Fabricius on day 11 (Leene et al., 1973). Thymus cells are present by day 11 and cell-mediated immunity has been demonstrated by day 13–14 (Solomon, 1971).

Burnet sustained that the first requirement of an adequate physiological theory of antibody production is "to account for the differentiation of function by which the natural entry of foreign microorganisms or artificial injection of foreign red cells provokes an immunological reaction while the physiologically similar autologous material is inert. (...) The 'self-pattern' is hereditary, while the process by which the self pattern becomes recognizable is acquired and takes place during embryonic or immediate past embryonic stages." (Burnet and Fenner, 1949).

As Burnet (1960) explained in his Nobel lecture: "my part in the discovery of acquired immunological tolerance was a very minor one, it was the formulation of an hypothesis that called for experiment. (...) I have introduced ideas about the evolution the process of self-recognition because a biologist I believe we know less about the process of differentiation and morphogenesis than about any other major field in biology. There is an insistent suggestion that immunological self-recognition is derived from the process by which morphological and functional intensity is maintained in large and long-lived multicellular organisms."

Medawar applied the technique of skin transplantation rather than by looking for the presence of absence of antibodies. If, to a new-born mouse of strain A, Medawar gave an adequate number of spleen cells from strain B mice, he found that a few weeks later the treated A mouse would retain indefinitely a skin graft of B which would otherwise have been rapidly rejected. This demonstration provided the proof for the validity of Burnet's conjectures. The theory postulates that for the successful transplantation of a healthy organ into a body containing a defective one, the normal immune mechanisms must be overcome so that the foreign organ will be "tolerated."

In a letter to Sexton (1991), Medawar wrote, "As a young research worker, trying with my colleague Rupert Billingham to make sense of the remarkable phenomenon that skin grafts exchanged between dizygotic twin cattle are accepted as readily as they are between monzygotic twins, I came across 'The production of antibodies' by Mac Burnet and Frank Fenner. This propounded the notion that later came to be called tolerance and made special reference to the work on cattle twins of Dr. Ray Owen in the Department of Agricultural Genetics in the University of Wisconsin. Owen's work enabled us to interpret our own findings on skin grafts in cattle and this in turn inspired us to provide the first demonstration of acquired immunological tolerance that provided the proof for the validity of Burnet's conjectures."

8.3 THE DISCOVERY OF ACQUIRED IMMUNOLOGIC TOLERANCE

Medawar's earlier research had focused on the rejection of skin grafts by burn patients (Medawar, 1946; Gibson and Medawar, 1943), using outbreed rabbits to investigate the process (Medawar, 1944, 1945, 1946; Gibson and Medawar, 1943). This research identified immune responses characterized by lymphocyte infiltration of genetically dissimilar grafts, but not of autografts, as being responsible for rejection in both species.

Billingham et al. (1953) published their first work in which they demonstrated that some mice inoculated in utero with a mixture of donor strain including splenocyte cells failed to reject donor strain skin grafts when these were transplanted 6 to 8 weeks after birth. In other words, donor splenocytes could be engrafted by their intravenous infusion into immunologically immature mice in utero or perinatally. When the inoculated recipients mature, they could accept skin and other tissues from the donor mouse strain.

Medawar predicted that an exchange of skin grafts between dizygotic calves would verify Burnet's hypothesis. Together with his postdoctoral fellow Rupert Billngham, he performed a series of grafting experiments that provided direct support for the concept of neonatally acquired transplantation tolerance. At the same time, Milan Hasek in Prague demonstrated that parabiosis of different strain chick embryos induced an immune hyporesponsive state to each other red cells.

Once the mice reached adulthood, Medawar performed skin homografts from the original donor strain. Transplantation of skin "was used for the study of a wide variety of biological problems, including the nature of pigmentation, the effect of freezing and drying upon the viability of tissues, the immunological effect of cortisone, and the role of different skin layers in engendering tumors under the influence of certain chemicals." (Billingham and Medawar, 1951).

The grafts were accepted, but grafts from third unrelated mouse strain were rejected, indicating that self was defined during embryonic development as Burnet had hypothesized (Billingham et al., 1953). Normally, mice reject skin grafts from other mice, but the inoculated mice in their experiment accepted the donor skin grafts. They did not develop an immunological reaction. The prenatal encounter had given the inoculated mice an acquired immunological tolerance. A significant proportion of the recipient became tolerant indefinitely. Medawar and his collaborators had proven Burnet's hypothesis. The immune system is not preprogramed to distinguish between self and nonself but learns to do so as a result of exposure to self-molecules during early development. In their paper, Medawar and his collaborators commented the previous findings of Owen (1945) as follows: "An exactly comparable phenomenon has been described by Owen, who found that the majority of dizygotic cattle twins are born with, and long retain, red blood cells belonging genetically to the zygote lineage of its twin...There is reason to doubt that this is because cattle twins, being synchorial, exchange blood in fetal life...." (Billingham et al., 1953).

As Park (2010) pointed out: "Medawar's experiments revealed more complex aspects of the tolerance phenomenon than Owens. It was found that 'the degree of tolerance' among the calves was 'widely variable'. While complete tolerance toward their dizygotic twin's skin was found in 36 cattles among the 42, the remaining six individuals also showed varied degrees of tolerance, measured by the number of days during which the skin patch survived on the host body. Moreover, it was found that the 'grafts from one twin to the other may be tolerated although grafts of the reciprocal transplantation are eventually destroyed'. These findings indicated that the tolerance was a highly complex phenomenon, and it was necessary for Medawar to study the varied dimension of the problem systematically using more standardized laboratory organisms, such as the mouse."

In 1953, the same year of the paper of Medawar's lab, a Czech immunologist, Hašek (1953) published a paper in which he described the technique of embryonic parabiosis, involving the anastomosis of two allogenic chick embryos through their chorioallantoic membranes, resulting in free exchange of fetal blood from about the 10th day of incubation. Through this technique Haŝek demonstrated that, after hatching, the parabionts were less able to form antibodies in response to each other's serum proteins. Haŝek met Medawar and Brent at an international embryology meeting and thereafter interpreted his data in terms of acquired tolerance. Billingham et al. (1956) published a second seminal paper including experiments showing that tolerance could be abolished in adult animals by the adoptive transfer of normal or presensitized lymphoid cells.

For the British immunologist Mitchison (1990), the work of Medawar on immunological tolerance: "performed the immensely important service of making transplantation scientifically respectable and gave the clinicians a well-defined goal to attain.

Moreover, antigen-specific suppression of the immune response by something akin the acquired tolerance remains an aim of research in transplantation and autoimmunity."

In his autobiography entitled "Memoir of a Thinking Radish," Medawar (1986) explained: "Thus the ultimate importance of the discovery of tolerance turned out to be not practical, but moral. It put new heart into the many biologists and surgeons who were working to make it possible to graft, for example, kidneys from one person to another."

In 1968, 15 years the paper of Billingham, Brent, and Medawar, Robert Good and Fritz Bach reported the first two successful human bone marrow transplants (Gatti et al., 1968; Bach, 1968), and in the 1980s, new immunosuppressive drugs were discovered, including cyclosporine, tacrolimus, sirolimus, and mTor inhibitors. As has outlined Park (2010), the work of Medawar and Burnet: "altered the direction of immunology's development from chemical studies of antigen-antibody reactions to biological investigations into living organisms and their continuously shifting physiological states."

REFERENCES

Bach, F.H., 1968. Bone marrow transplantation in a patient with the Wiskott–Aldrich syndrome. Lancet. 2, 1364–1366.

Billingham, R.E., Medawar, P.B., 1951. The technique of free skin grafting in mammals. J. Exp. Biol. 28, 385–402.

Billingham, R.E., Brent, L., Medawar, P.B., 1953. Actively acquired tolerance of foreign cells. Nature. 172, 603–606.

Billingham, R.E., Brent, L., Medawar, P.B., 1956. Quantitative studies on tissue transplantation immunity. III. Actively acquired tolerance. Philos. Trans. R. Soc. Lond. B 239, 357–414.

Brent, L., 1997. The discovery of immunologic tolerance. Hum. Immunol. 52, 75–81.

Burnet, F.M., 1960. Immunological recognition of self. Nobel Lect. 659–701.

Burnet, F.M., 1971. Walter and Eliza Hall Institute 1915–65. Melbourne University Press, Melbourne, p. 68.

Burnet, F.M., Fenner, F., 1949. The Production of Antibodies. Macmillan, Melbourne, Australia, p. 86.

Gatti, R.A., Meuwissen, H.J., Allen, H.D., et al., 1968. Immunological reconstitution of sex-linked lymphopenic immunological deficiency. Lancet. 2, 1366–1369.

Gibson, T., Medawar, P.B., 1943. The fate of skin homografts in man. J. Anat. 77, 299–310.

Hašek, M., 1953. Vegetative hybridization of animals by means of junction of the blood circulation of the blood during embryonic development. Cesk. Biol. 2, 267–282.

Leene, W., Duyzings, M.J.M., Von Steeg, G., 1973. Lymphoid stem cell identification in the developing thymus and bursa of Fabricius of the chick. Z. Zellforsch. 136, 521–533.

Medawar, P., 1986. Memoir of a Thinking Radish: An Autobiography. Oxford University Press, Oxford.

Medawar, P.B., 1944. The behavior and fate of skin autografts and skin homografts in rabbits: a report to the War Wounds Committee of the Medical Research Council. J. Anat. 78, 176–199.

Medawar, P.B., 1945. A second study of the behaviour and fate of skin homografts in rabbits: a report to the War Wounds Committee of the Medical Research Council. J. Anat. 79, 157–176.

Medawar, P.B., 1946. Immunity to homologous grafted skin; the suppression of cell division in grafts transplanted to immunized animals. Br. J. Exp. Pathol. 27, 9–14.

Mitchison, N.A., 1990. On P.B. Medawar 1915–1987, elected FRS 1949. Biographical Mem. Fellows R. Soc. 35, 283–301.

Murphy, J.B., 1913. Transplantability of tissues to the embryo of foreign species. Its bearing on questions of tissue specificity and tumor immunity. J. Exp. Med. 17, 482–493.

Owen, R.D., 1945. Immunological consequence of vascular anastomoses between cattle twins. Science 102, 400–405.

Park, H.W., 2010. "The shape of the human being as a function of time": time, transplantation, and tolerance in Peter Brian Medawar's research, 1937–1956. Endeavour. 34, 112–121.

Sexton, C., 1991. The Seeds of Time. The Life of Sir Macfarlane Burnet. Oxford, Oxford University Press, Australia, p. 136.

Solomon, J.B., 1971. Lymphocytopoiesis and ontogeny of defined in birds. Fetal and Neonatal Immunology, Frontiers of Biology. Plenum Press, New York, Monograph 20.

CHAPTER 9

The Saga of Mast Cells: From Paul Ehrlich Doctoral Thesis to the Discovery of IgE

9.1 PAUL EHRLICH'S DISCOVERY

Mast cells (MCs) are bone marrow-derived tissue-homing leukocytes that participate to both innate and adaptive immunities. The history of MC research begins with a name and a date. The name is that of Paul Ehrlich (Fig. 9.1); the date is June 17, 1878. That day, the 24-year-old medical student from Strehlen (Schlesien) presented his doctoral thesis at the Medical Faculty of Leipzig University. The title of his dissertation was "Beiträge zur Theorie und Praxis der histologischen Färbung" (Contribution to the theory and practice of histological dyes) (Ehrlich, 1878) (Fig. 9.1).

This work is a beautiful and admirable example of analytical experimental method and foresight. Ehrlich's thesis was organized into two parts. In the first part, he overviews the chemical bases of many important histological reactions, and in the second, he discusses the chemical, technological, and histological properties of aniline dyes. In the chapter dedicated to the histological applications of this class of chemical compounds, he presents his personal point of view about a type of cell, which he named as "MC" ("Mastzelle"). This name, which means "well-fed cells" was attached to the newly described cell population in the belief that their aniline-positive metachromatic granules might contain deposits of nutrients and might developed as a result of hypernutrition. He was likely to be mistaken about this issue, although it should be noted that observations in MC-deficient mice suggest roles for MCs in control of diet-induced obesity (Liu et al., 2009).

Ehrlich stated, "aniline dyes display an absolutely characteristic behavior toward the protoplasmic deposits of certain cells" that were "chemically so sharply" distinguished from the group of Waldeyer's "Plasmazellen." With this term, he referred to a broad and heterogeneous category of cells previously described by Waldeyer. Among Waldeyer's "Plasmazellen," there was a group of connective tissue cells

Milestones in Immunology. DOI: http://dx.doi.org/10.1016/B978-0-12-811313-4.00009-7

Figure 9.1 A portrait of Paul Ehrlich.

exhibiting large dimension and round shape, which could be distinguished "from white blood cells on the basis of their significantly large size and lack of contractile activity." Ehrlich emphasized his assumption that most of those cells that he had described in connective tissues as reactive to aniline staining did not correspond to Waldeyer's description, which was otherwise based on purely morphological criteria, not chemical. "Anilophilic cells should be strongly separated from 'Plasmazellen'," he insisted. These aniline-reactive cells "represent sui generis elements and must be distinguished from Waldeyer's 'Plasmazellen' by a different denomination."

Ehrlich then came to the central part of his presentation. "From the descriptive point of view," he said, aniline-positive cells should be "most conveniently described as 'granular cells of the connective tissue'; from the physiological standpoint, these cells may provisionally be indicated as MCs ('Mastzellen') because, like fat cells, they represent a further development of the fixed cells of connective tissue." Ehrlich's concept is absolutely remarkable in that, although MCs "are localized with extremely high frequency around blood vessels in the loose connective tissues," "it seems not justified to regard them as members of a perivascular system." He also provided a notable explanation to support his view: aniline-reactive cells indeed "have a tendency to collect around developing preformed structures in connective tissues." In discussing this point, he added, "in certain acinar glands (goat parotid), the pattern of MC accumulation [inside the organ] is not determined by the branching of the vascular system but by the ramification of the gland excretory ducts."

In the course of his dissertation, Ehrlich underlined the concept that MCs must be principally distinguished on the ground of their reactivity to aniline dyes, not simply by their shape and morphological appearance. "Granular cells are characterized by the presence of a still undetermined chemical substance," "which is bound to the granular storages in the protoplasm" and which reacts to aniline dyes giving typical metachromasia. The binding of this chemical substance to aniline dyes shows different staining: red–violet with cyanine, orange with fuchsin and red with dahlia and gentian. Finally, he provided an extremely precise description of MC microscopical features. "The typical aspect of 'granular cells' is as follows. The mostly stainless protoplasm appears as being filled by more or less numerous grains of variable size. These granules exhibit subtle nuances specific for each staining procedure. The nucleus is mostly not stainable, even in samples which otherwise display beautiful nuclear staining reactions. In flattened cells, the nucleus appears as a characteristically clear spot, due to the absence of the colored granules and this picture nearly gives the impression of a lacuna in the cell body."

Many interesting aspects of Ehrlich's dissertation deserve some comments. He first coined the term "Mastzellen" to describe the aniline-reactive granular cells he found in connective tissues. The German word "Mast" implies a nourishing and "suckling" function for these cells. Certainly, MCs do not provide nutrients in a strict sense; however, they are deeply involved in the "trophism" of tissues. MCs are increasingly being recognized as key cells for connective tissue homeostasis, remodeling and repair. They also express relevant angiogenic activity. Their granules indeed contain proteases and cytokines that are known to exert "trophic" effects (survival, growth and chemotactic) on different cells, such as fibroblasts, myofibroblasts, smooth muscle cells, neurons and endothelial cells. Therefore, the "provisional" term "Mastzellen" seems more and more appropriate for describing these cells. Ehrlich also observed that MCs did not strictly belong to a diffuse perivascular system, despite their characteristic arrangement close to capillaries. This is indeed an absolutely correct statement.

MCs often localize far from blood vessels and also express a series of biological properties that are not related to microvascular functions. He argued that MCs could also be found around areas of developing tissues. The close relationship between MCs and tumor growth is of extreme actual interest in the sense proposed by Ehrlich. In addition, he pointed out that the use of aniline dyes was of the utmost importance for identifying MCs. Reactivity of aniline with a "still undetermined chemical substance" stored in the granules was the sole reliable procedure that would enable the microscopist to recognize these cells with certainty. We now know that aniline dyes interact with the highly acidic glycosaminoglycan residues contained within MC granules. This reaction, in turn, determines the characteristic metachromasia of such structures. We acknowledge that his advice, not simply to consider cell morphology but to base cell identification upon a specific histochemical reaction, was an extremely modern concept. As to the origin of MCs, we now know that they do not differentiate from fibroblasts, as suggested by Ehrlich. He could not imagine, however, that these cells would derive from precursors of the hematopoietic lineage and complete their differentiation in peripheral tissues. This was certainly more than he could determine with the simple support of a light microscope and some histological dyes.

On January 17, 1879, the Physiological Society of Berlin heard a remarkable paper by Paul Ehrlich about the MCs that he had discovered as a medical student two years previously. Ehrlich pointed out that not only do the granules of mammalian MCs display great avidity for basic dyes, but that they also tend to alter the shade of the dye (metachromasia). Later (with one his pupils), he stressed a second characteristic feature of the MC granules in many species, their solubility in water (Westphal, 1891). Michels (1938) wrote, "uncounted pages of useless and misleading research have been the result of the failure on the part of many investigators to heed the admonition originally given by Ehrlich and Westphal, that the MC granules are soluble in water and that to preserve them tissues must be fixed in 50% alcohol and stained in alcoholic thionine." Ehrlich then went on to study the staining reactions of blood cells, laying the foundations of modern hematology on the basis of the specific affinities of the leukocytes for various dyes (Ehrlich, 1879; Ehrlich and Lazarus, 1898). He encountered cells with basophilic, metachromatic granules and, thus, came to recognize two types of MCs: the first—derived from, and living in, the connective tissues (tissue MCs); the second—the counterpart of the neutrophil polymorph and eosinophil leukocyte—with its origin in the bone marrow and habitat in the peripheral blood (blood MC, basophil or mast leukocyte). Meanwhile, Ehrlich (1879) had discovered basophilic granular cells in human blood, although so far only in myeloid leukemia. Nevertheless, with characteristic insight, he at once perceived that, in higher vertebrates, the blood MCs are true leukocytes stemming from precursors in the bone marrow. By the time that his textbook (Ehrlich and Lazarus, 1898) was revised in 1909, the evidence for the myeloid origin of the blood MC was complete (Jolly, 1900).

Later work established that MCs and basophils share several notable features besides staining properties. Both cell types represent a major source of histamine and other potent chemical mediators implicated in a wide variety of inflammatory and immunological processes. To study the presence and significance of MCs in pathological conditions again acknowledges our debt to the pioneer observations of Ehrlich who described two situations in which connective tissue may be overnourished, in chronic inflammation and the microenvironment of tumors. Here, there exists a lymph stasis, a damming up of tissue fluid rich in nutriment, whereby certain fixed connective tissue cells are stimulated to become mobile, to multiply, and to convert some of the abundant extracellular material into specific intracellular granules. According to Ehrlich, MCs were "indices of the nutritional state of the connective tissue," increasing during periods of hypernutrition, diminishing during periods of relative starvation. Ehrlich found many MCs in tumors, especially carcinoma, but it was left to his pupil to recognize that the cells tend to accumulate at the periphery of carcinomatous nodules rather than within the substance of the tumor (Westphal, 1891). The number of MCs within the perivascular and interstitial connective tissue of different neoplasias has been reported to be increased. In some cases, this phenomenon is a characteristic feature of the lesion.

9.2 FURTHER EVIDENCE AFTER EHRLICH'S DISCOVERY AND THE FUNDAMENTAL CONTRIBUTION OF WILLIAM BATE HARDY

Sixteen years after Ehrlich's first description of Mastzellen, the English histologist and physiologist William Bate Hardy (Fig. 9.2) provided a further contribution to the histochemical and functional definition of MCs. In the beginning, he referred to the formerly described Ehrlich's Mastzell with the collective term of "coarsely granular basophile cell" (CGBC), but in two outstanding papers, published in the Journal of Physiology in the years 1894 and 1895 (Hardy and Wesbrook, 1895), he distinguished two types of granular basophile cells, i.e., the CGBCs and the "splanchnic basophile cells," which both belonged to the population of "wandering cells" (the modern leukocytes). Hardy was born on April 6, 1864 at Erdington, Warwickshire. He went to Gonville and Caius College, Cambridge, where he was awarded his Bachelor of Arts degree in 1888. He had a versatile mind and soon became a renewed histologist and physiologist. He was attracted by different scientific themes, from zoological subjects to the structure of the cell protoplasm. He laid the foundation of the modern knowledge of colloidal electrolytes, showing protein migration in an electric field and establishing the existence of the "isoelectric point." Hardy's brilliant competence was recognized by the scientific community. He was elected Vice President of the Royal Society from 1914 to 1915 and Secretary from 1915 to 1919. He was also knighted in the New Year's Honors of 1925. Hardy lived and worked mostly in the School of Physiology in Cambridge. He died on January 23, 1934 and was buried at Caius College Chapel, Cambridge.

Hardy studied the staining characteristics and functional properties of the "wandering cells"—with this term, the authors intended cells endowed with the power of ameboid movement—in the coelomic cavities and the connective tissue of different mammalian species. Hardy referred to the formerly described Ehrlich's Mastzell with the term of CGBCs. He recognized that these cells stained with methylene blue solution or with glycerin eosine and Loeffler's methylene blue. He also performed functional studies by producing localized cell accumulation in connective tissues, using chambers or fine tubes filled with bacilli or their products or chemical irritants, and placed under the skin, or in a coelomic cavity. In addition, he exploited the technique of the introduction of copper into the anterior chamber of animal eyes and the production of skin blisters on himself or on others by an irritant. Hardy was aware of the fact that CGBCs from the coelomic spaces and from connective tissues differed slightly. He stated, "The cells [the CGBCs] of the connective tissues and those of the coelomic fluid differ slightly in size and shape, and we will therefore deal with them

Figure 9.2 A portrait of William Bate Hardy.

separately." He also recognized that the CGBC was "not strictly a wandering cell" but a "stationary" cell "except perhaps under certain conditions," being thus somewhat dissimilar from the other blood- and lymph-borne wandering cells. As to the staining properties of CGBC granules, Hardy wrote illuminating words, "When microbes or microbic poison or 'irritants' are present, these cells are frequently found with their granules so changed that they no longer stain, or stain imperfectly. In place of all the granules in the cell staining, all or some of the granules are now refringent spherules of the same size as the normal granules, but they either completely refuse to stain, or they stain in patches."

Here, Hardy was describing the phenomenon of MC degranulation, which left completely or partially emptied granule containers. The interpretation he gave was outstanding, "The phenomenon very strongly suggests that these granules are not entirely composed of the basophil substance, but rather are composed of an unstaining groundwork with which the basophil material is associated but from which it may be removed." Thus, it appears that Hardy had correctly recognized the heterogeneous nature of MC granules. Today, we know that these organelles are complex structures which store a large series of biologically active compounds, e.g., histamine (also serotonin in rodents and fish), some preformed cytokines and growth factors, as well as many proteases embedded in a glycosaminoglycan meshwork (Galli et al., 2005).

Remarkably, he underlined that the "unstable" or "explosive nature" (in italics in the original text) of the CGBCs in certain animals, like guinea pigs and rabbits, was "one of their most remarkable characters," adding, "under the influence of certain chemical stimuli, [...] the basophil cells of the rat also become explosive." Thus, CGBCs were heterogeneous cells in so far as cells isolated from different animals exhibited different propensity to release ("explode") their granule constituents. He noted that this event caused the complete, although apparent, disappearance of CGBCs under the microscope. Here, Hardy was referring not only to the intrinsic fragility of some MC types but also to their susceptibility to undergoing generalized exocytosis, which leaves cellular ghosts.

Interestingly, 60 years later, another great scientist in MC research, James F. Riley, who discovered histamine in these cells, spoke about "exploding" MCs that were visible under the microscope when an electric current was passed on a fresh spread of rat mesentery (Riley, 1955). As to the tissue distribution of these cells, Hardy recognized that, unlike the finely granular basophil cell (the modern basophil), CGBCs occur only in the extravascular spaces and "are not merely rare but completely absent from the blood." In a remarkable footnote, he wrote, "blood obtained by plunging a fine pipette into the heart rarely fails to contain one or two [CGBCs] removed from the pericardial space. Blood from a vessel is always entirely free from them." In experiments with Ziegler's chambers filled by bacteria and placed under the skin, Hardy observed that the connective tissue that formed the wall of the implantation cavity "was packed with an immense number of the coarsely granular oxyphil cells [the modern neutrophils] together with a smaller number of CGBCs," a discovery which in some way forecasts the notion of the MC as a tissue sensory cell that accumulates at site of tissue injury and recruits inflammatory elements, in particular neutrophilic leukocytes. He also found that introduction of cultures of bacilli into rat or guinea pig peritoneal cavities caused instantaneous disintegration of a considerable number of CGBCs, so that extensive free basophilic granules could be detected in the peritoneal fluids.

Remarkably, these findings imply triggering the process of massive exocytosis which occurs in peritoneal MCs upon stimulation. In the second published paper (Hardy and Wesbrook, 1895), Hardy went yet deeper into the concept of MC heterogeneity. Here, he examined the wandering cells "which lie in the interspaces of the mucous coat of the gut" in a number of mammalian, reptilian and amphibian species. Wandering cells of the alimentary canal were collectively referred to as "splanchnic cells." The article opens with a technical note. Hardy pointed out, "the basophil cells of the gut are always exceedingly sensitive to the presence of even minute traces of water" in the fixative solutions, suggesting the absolute necessity of alcohol solutions with no less than 80% alcohol present as the only reagents which preserved this kind of cells. He noted that there was heterogeneity of basophilic cells as to their number and chemical properties in the different species examined. He found that splanchnic basophil cells "are especially striking in Carnivora both as regards their number [...] and the orderly arrangement they exhibit. In Herbivora on the other hand these cells are both less numerous and less ordered." He also quantified the population of basophil cells in the dog intestine, concluding, "as many as three to five hundred are present in a single villus." Staining differences were also recognized among basophil cells belonging to different species. He wrote, "slight differences in the degree of instability of the granules of the splanchnic basophil cells were found in different animals; those of the rat for instance being a little more resistant than those of the rabbit." He also added, "the differences between the splanchnic and coelomic basophil cells of the rat are exceedingly striking both as regards the size and stability of the granules and the size of the cells." All this information came 70 years before Enarback's differentiation between connective tissue-type and mucous-type MCs in rodent species (Enerback, 1966a,b, 1986).

Most remarkably, Hardy added, "it should be noted that the splanchnic basophil cells occur only in the mucous coat of the gut, and a section treated with water will still show CGBCs of the ordinary type in the submucous and muscular coats." An outstanding sentence, indeed, which established the basically different histochemical properties of the mucosal-type MCs and the connective tissue-type MCs ("the ordinary type" cells). And, in addition, "in Mammalia the splanchnic basophil cell always differ from the CGBCs (coelomic) in being smaller and possessing usually very much smaller granules." Thus, not only a staining heterogeneity could be documented among MCs but also a structural one. "Spanchnic basophil cells themselves vary much in shape [...] they may appear elongated and flattened or rounded and lobed." In the villi of Carnivora, Hardy found a remarkable arrangement of spanchnic basophil cells. They were "flattened in the plane of the surface of the villus" and formed a sort of continuous subepithelial layer. Other spanchnic basophil cells lay scattered in the axis of the villus. He wrote that, proceeding downwards into the crypt region, these scattered cells "become more and more numerous so that a considerable number lie between the crypts and in the basal region." The identification of a subepithelial layer of spanchnic basophil cells in the intestinal villus is remarkable and brings to mind recent publications concerning the potential of MCs as important factors in affecting villus architecture (Crivellato et al., 2005, 2006).

Although, in Hardy's opinion, splanchnic basophil cells belonged to wandering cells, their migration was to be considered a rare occurrence. Their movements indeed were mostly limited to changes of the cellular shape. He wrote, "the fixed nature of these cells is shown by the fact that, though the numbers present do vary in different animals, even in different individuals of the same species, yet we have never met with any increase or decrease in number sufficient to warrant us in thinking that they commonly vary very greatly." He further clarified his concept in the following way, "we were for a long time under the impression that these splanchnic basophil cells were not in the strictest sense wandering cells, we supposed that their movements were limited to changes of shape of the cell body and did not carry the cell from place to place. This idea, however, must be given up." In two cases, indeed, he witnessed a real cell mobilization of this kind of cells in so far as "undoubted basophil cells were found thrust between the cells of the endodermic epithelium." Remarkably, in one of these cases "the small intestine contained very large numbers of bacilli." Thus, he recognized that intestinal MCs may migrate and acquire an intraepithelial position as a consequence of strong bacterial invasion. One of the most intriguing issues discussed in this paper concerned the chemical nature of basophil granules. Hardy wrote, "though nothing is known of the chemical nature of these granules yet there is reason for believing that they are not simply proteid in nature." As previously mentioned, he recognized that they were unstable in presence of water even after prolonged exposure to absolute alcohol, a procedure that obviously fixed proteic structures but not nonproteic constituents, such as biogenic amines. Hardy's view of basophil cell function was partly in line with Ehrlich's concept of a nutritional role for these cells. Hardy performed a series of experiments on starving and hyper-fed animals, finding that splanchnic basophil cells changed the morphology and histochemical property of their granules in parallel to changes in the nourishing regimen. He claimed, "the cells are markedly granular in well-nourished animals and become less granular during starvation." However, and most remarkably, he recognized, "the latter change is not readably brought about and it is difficult to produce extreme exhaustion of the granules." He also seemed to link basophil cell function to inflammation. He wrote, indeed, "hypertrophy of basophil cells [...] commonly occurs when abnormal chemical substances are present, e.g. during inflammation." He also suggested that these cells may provide phagocytic or clearing activity under certain conditions. He stated, indeed, "the basophil cells remove certain substances." This was suggested, in particular, by the ordered subepithelial arrangement found in Carnivora. Here, the splanchnic basophil cell lies in an ideal position to absorb the fluid elaborated by the activity of the endodermic epithelium before it reaches the blood vessels and lymphatics. Thus, in Hardy's opinion, the basophil cell may "take part in the manipulation of the absorbed products of digestion," although he correctly recognized that these cells were found "in a fetus at a period of life when the gut is not occupied in the digestion of food."

Controversies arose but their resolution for the most part merely emphasized the soundness of Ehrlich's original work. However, the functional biology of MCs resisted clarification until recently, as their role in promoting the nonspecific inflammatory reaction and in different immune responses could be elucidated. Also, neoplasias arising from MCs have been elusive to clinicians, hematologists, and pathologists. The origin of MCs remained obscure for many years. It is now accepted that MCs arise from pluripotential hematopoietic cells in the bone marrow that express CD34, c-kit and CD13 (Kirschenbaum et al., 1991). This was demonstrated for the first time by Kitamura et al. (1978), who performed in vivo experiments using genetically

MC-deficient mutant mice. However, in contrast to other cells of the hematopoietic stem cell lineage, which differentiate in the bone marrow before being released into the circulation, MCs do not circulate as mature cells, but in small numbers as committed progenitors. The progenitors complete their maturation with concomitant phenotypical diversity after moving into diverse peripheral tissues. The concept of "MC heterogeneity" has represented a focal point in recent discussions of MC biology, and it emphasizes that different MC populations exhibit significant variation in multiple, potentially important aspects of their phenotype. MCs from different species, from different sites in the same species and even from the same organ in one species can vary in their response to stimuli and inhibitors of mediator release.

Observations of histochemical and functional heterogeneity of MCs, first given a sound basis in the 1960s by Enerback (1966a,b, 1986), are received increasing attention (Galli, 1990). Enerback reported that, in contrast to MCs in rat skin, MCs in the intestinal mucosa were sensitive to routine formalin fixation and could not be identified in standard histological sections. However, after appropriate fixation and sequential staining with Alcian blue and safranin, the mucosal MCs stained blue in comparison with the connective tissue MCs, which stained with safranin and were red. There is no disease, biological condition or animal model yet identified that exhibits an absolute lack of MCs from which or in which their biological role might be inferred. MCs are most commonly regarded as key effectors in the pathogenesis of allergic diseases. However, an exciting development in the study of MC biology was the discovery that MCs can generate or release various cytokines, which indicate a key role played by MCs also in diverse pathophysiological processes, such as chronic inflammatory processes, wound healing, angiogenesis, fibrosis and tumors. All scientists involved in the field of MC research should acknowledge their debt to Ehrlich's pioneering observations.

These tissue-homing cells corresponded to the subsets of connective tissue-type and mucosal MCs, respectively, which would be described seventy years later by Enerbäck in rodents (Enerback, 1966a,b). Among the CGBCs, he also differentiated those cells which populated the serosal cavities—the so-called coelomic CGBC—from the common CGBCs which were localized in the connective tissues. He stated that the granular basophile cells presented with different morphological and histochemical characteristics in diverse animal species as well as at different anatomical sites, being thus the first scholar to shape the fundamental concept of "MC heterogeneity."

MCs were demonstrated in most animal species, although with an irregular, capricious distribution (Michels, 1938). In his for long unsurpassed review of the MCs, Michels praised Ehrlich's pioneering contribution to the study of these cells, in particular his recognition that MCs granules were soluble in water. He wrote, "uncounted pages of useless and misleading research have been the result of the failure on the part of many investigators to heed the admonition originally given by Ehrlich and Westphal, that the MC granules are soluble in water and that to preserve them tissues must be fixed in 50 per cent alcohol and stained in alcoholic thionine" (Michels, 1938).

As for the relation of MCs with basophils, after an initial unitary conception, subsequent studies indicated that at least in higher organisms these two cells differed both in habitat and in parentage, being the derivation of MCs unknown—they were usually interpreted as istiogenic elements—whilst the origin of basophils was from the bone marrow. Michels wrote, "aside from an identical basophilic metachromatic reaction of the granules, the two cell types have nothing in common" (Michels, 1938).

9.3 HEPARIN, HISTAMINE, AND SEROTONIN CONTAINED IN MAST CELL SECRETORY GRANULES

At the end of the 1930s, a group of Scandinavian researchers provided fundamental new insight as to MC structural and functional profiles. The mysterious MC component prophesized by Ehrlich as the responsible agent for granule metachromasia was revealed by Jorpes, Holmgren, and Wilander (Holmgren and Wilander, 1937; Jorpes et al., 1937). Following Jorpes' discovery that the anticoagulant heparin—a polysulphuric acid ester, made up of glucuronic acid, glucosamine and sulphuric acid—was subject to stain metachromatically with toluidine blue, Holmgren and Wilander reconsidered Ehrlich's observation that MC granules stained metachromatically with toluidine blue. These authors were able to set a correlation between the number of toluidine blue-positive MCs in various tissues and their heparin content. Tissues with large amounts of "Ehrlichschen Mastzellen" were particularly rich in heparin and, among MC-rich tissues, the beef liver capsule was described as "a pure culture of MCs" (Holmgren and Wilander, 1937; Jorpes et al., 1937). The Swedish investigators formulated the conclusive theory that the task of MCs in the connective tissues was to produce heparin.

The discovery that tissue MCs were the source of heparin was the prelude to the identification of two other crucial substances contained in MCs: histamine and serotonin. A potential correlation between tissue heparin and tissue histamine contents was initially established. Indeed, the release of histamine was shown to be accompanied by a similar release of large amounts of heparin both in vivo and in vitro (Rocha and Silva, 1942).

In a series of fundamental studies published in the period 1952–1956, the pharmacologists Riley and West in Scotland demonstrated that histamine—the previously identified Lewis' "H substance" responsible for skin anaphylactic phenomena—was present in MCs. Early studies showed that injection in the rat of histamine liberators, such as stilbamidine and D-tubocumarine, was followed by selective damage to tissue MCs indicating that MCs were the presumptive site of histamine accumulation in the tissues (Riley, 1954).

Further investigations revealed that very high values for heparin and histamine could be found in tissues which were exceptionally rich in MCs, such as the cleaned capsule of normal ox liver, and the sheep and ox pleura (Riley, 1955). The loose connective tissues but not the dense connective tissue of the tendons were rich in histamine and MCs as well. A strong positive correlation between histamine tissue contents and the histological demonstration of MCs was also recognized in pathological conditions such as urticaria pigmentosa in man (Riley and West, 1953) and MC tumors in dogs (Cass et al., 1954).

Further evidence for the presence of histamine in MCs was provided by Fawcett and by Mota and Vugman. Fawcett (1954) demonstrated that the potent histamine liberator, compound 48/80, caused release of MC granules, and that it failed to liberate appreciable amounts of histamine from connective tissue previously depleted of MCs.

Mota and Vugman (1956) reported a good correlation between serum histamine levels and disruption of MCs in a guinea pig model of anaphylaxis. On the whole, these data show that by the end of the 1950s, experimental studies had delineated a fundamental functional link between MCs, histamine, and the allergic and anaphylactic reactions which had been recognized and described long before by Pirquet, who coined the term "allergy" in 1906, and Paul Portier and Charles Richet, who introduced the term "anaphylaxis" in 1902.

Thus, MCs could since be defined as the major tissue repository for histamine ("histaminocytes") and were entitled to play a crucial role in allergic conditions such as hay fever, asthma and anaphylactic shock (Beaven, 2009). It soon appeared, however, that basophils too were rich in histamine and heparin (Graham et al., 1955; Beherens and Taubert, 1952). Thus, the similarities between MCs and basophils seemed again to outweigh their differences (Riley, 1954).

About the same period, Benditt et al. (1955) demonstrated that 5-hydroxytryptamine (5-HT, serotonin) associated with MCs of the subcutaneous areolar tissue of the rat. Serotonin was a vasoconstrictor substance suspected for decades to be contained in platelets. It was isolated and characterized in 1948 by Maurice Rapport and Irvine Page and discovered to correspond to enteramine by the Italian scientist Vittorio Erspamer in 1952 (Rapport et al., 1948; Erspamer and Asero, 1952).

Later work by Parratt and West (1957) revealed that serotonin was concentrated in tissue MCs of the rat and the mouse but not of the guinea pig, dog, man, rabbit, cow, hamster, and cat, and that the skin of the rat contained more than half of the total serotonin of the body. These authors speculated that MC serotonin might be involved in the response of animals to injections of large-molecular-weight substances such as egg-white or dextran. Both histamine and serotonin had potent effects, especially on the vascular system, and their release by MCs, which expressed a preferential perivascular location, could be implicated in the inflammatory reactions occurring in connective tissues (Riley, 1963). Remarkably, not only MCs had the ability to synthesize histamine and serotonin but, as demonstrated by Green's group, both normal and neoplastic MCs were able to take-up exogenous histamine and serotonin (Day and Green, 1962; Furano and Green, 1964). It was thus made clear that MCs can concentrate biogenic amines.

9.4 MAST CELL HETEROGENEITY

Hardy's concept of MC heterogeneity was further developed in the 1960s by Lennart Enerback. Based on their specific staining characteristics and preferential tissue homing, two morphologically distinct subpopulations of rodent MCs were initially identified and termed connective tissue MCs (CTMCs) and mucosal MCs (MMCs), respectively (Enerback, 1966a,b, 1986). The former populated the mucosae of the respiratory and gastrointestinal tracts, while the latter homed to the connective tissues and serosae. CTMCs could be distinguished from MMCs by staining in red with safranin due to the presence of large amounts of heparin in their secretory granules.

In the mouse, indeed, the proteoglycan content of MC granules varied in the different MC subtypes. CTMCs contained heparin that lacked in MMCs. Conversely, MMCs expressed chondroitin sulfates A and B, which were not found in CTMCs, whereas both MC subtypes stored chondroitin sulfate E in their granules. Thus, in contrast to CTMCs, MMCs were sensitive to routine formalin fixation and could not be identified in standard histological sections. After appropriate fixation and sequential staining with Alcian blue and safranin, the MMCs stained blue, being thus differentiated from CTMCs which stained with safranin and were red. It later appeared that CTMC and MMC subtypes contained distinct classes of proteases and expressed different functional profiles, being activated in part by different stimulators and providing selective secretory responses.

9.5 THE DISCOVERY OF IGE

During the 1970s, other crucial aspects of MC involvement in allergic and anaphylactic reactions were recognized. It was not until 1967 that the "reaginic" antibody—the transferable factor responsible for the sensitization phenomenon described by Prausnitz in 1921 (Prausnitz and Küstner, 1921)—was eventually identified by Kimishige Ishizaka and Teruko Ishizaka (Fig. 9.3) as γE-antibodies (immunoglobulin E, IgE), a minor component of the immunoglobulin family (Ishizaka and Ishizaka, 1967). IgE was shown to be capable to mediate the release of histamine and another mysterious substance called "slow reacting substance of anaphylaxis" (SRS-A) from sensitized tissue MCs (Ishizaka et al., 1970; Orange et al., 1971).

The receptor for IgE molecules was later identified at high concentrations on the surface of MCs and was recognized to bind IgE with high affinity and specificity. This was named the "high affinity" receptor for

Figure 9.3 A portrait of Kimishige Ishizaka and his wife Teruko.

IgE (FcεRI), and it was completely cloned in 1989 (Blank et al., 1989). The crosslinking of IgE with bivalent or multivalent antigen on the surface of MCs resulted in the aggregation of IgE receptors and in the triggering of MC degranulation. SRS-A was initially recognized by Feldberg and Kellaway (1983) as a spasmogenic substance distinct from histamine. This stuff was capable to increase microvascular permeability and produce long-lasting wheal-and-flare responses in the skin and bronchoconstriction in the lungs during anaphylactic shock (Brocklehurst, 1960). It became apparent that SRS-A was a mixture of lipids, collectively named leukotrienes (LTs), which were generated from arachidonic acid through the 5′-lipoxygenase pathway. MCs were recognized to synthesize LTC_4, LTD_4, and LTB_4 as well as other inflammatory lipids (eicosanoids) called prostaglandins (PGDs) (Roberts et al., 1979). Within minutes of MC stimulation with anti-IgE antibodies, which activate the IgE receptor on the cell surface and trigger MC response, MCs were seen to release substantial amounts of PGD_2 (Lewis et al., 1982).

Parallel investigations showed that basophils were also endowed with the high-affinity receptor for IgE, a discovery that further linked the two cell lineages in the speculative approach of researchers. In addition, both MCs and eosinophils were recognized to participate to certain protective reactions against parasites (Askenase, 1977). Thus, until the mid-1990s, the paradigm prevailed that MCs were to be regarded as tissue elements principally implicated in the pathogenesis of allergic reactions and responsible for defense against certain parasites.

9.6 ULTRASTRUCTURAL EVIDENCE

In the 1970s, accurate definition by transmission electron microscopy of a series of MC structural and functional details was provided by several groups. In particular, the group of Ann Dvorak in Boston afforded persuasive demonstration of the heterogeneous structure of MC granules, clarifying the mechanisms of MC degranulation and recovery, and providing new data on the fine, distinctive aspects of MC and basophils ultrastructure.

During MC degranulation, cytoplasmic granule membranes were seen to fuse with each other and with the plasma membrane, giving rise to open secretory channels which allowed the release of granule contents into the extracellular environment (Dvorak and Kissel, 1991). This quick, explosive, IgE-mediated process of MC degranulation, characteristic of type I hypersensitivity reactions and categorized as "anaphylactic degranulation" or "compound exocytosis," was punctually differentiated by a novel pattern of MC secretion Dvorak's group was able to identify, for which the term "piecemeal degranulation" was coined.

In a series of elegant ultrastructural immunocytochemical experiments, Ann Dvorak and coworkers described the fine aspects of this newly identified degranulation pathway, which allowed MCs to release subtle amounts of granule-stored material in a prolonged time lapse. There was a slow discharge of granule contents in a "piecemeal" fashion, without membrane fusion events and granule opening to the cell exterior. Dvorak underlined the concept, "piecemeal degranulation" represented the most common way of MC secretion observed in MCs infiltrating areas of chronic inflammation or tumors whilst the well studied and much more renowned pattern of "anaphylactic degranulation" could rarely be recognized apart from the sites of allergic responses (Dvorak and Kissel, 1991).

The primary involvement of MCs in such so harmful and sometimes life-threatening events as allergic and anaphylactic reactions left researchers somewhat disconcerted as to the real physiological role of MCs. How was it possible that so a ubiquitous and universally distributed cell type might exist only to cause danger to the host?

Still in 1975, Harold Dvorak and Ann Dvorak wrote, "much more must be learned before we can confidently describe the role of basophils, or of the closely related MCs, in health or disease. It seems most unlikely that either cell exists for the purpose of destroying the organism in anaphylactic shock. Nonetheless, it is highly probably that basophil/MC function is closely related to the potent chemicals stored within their cytoplasmic granules. One likely possibility holds that small amounts of these chemicals are required for homeostasis (e.g., for regulation of the tone of the microvasculature) and that these cells function by releasing such substances continuously, as they are needed, in small aliquots rather than by explosive discharge" (Dvorak and Dvorak, 1975).

9.7 THE ORIGIN OF MAST CELLS

By the end of the 1970s, scientists were able to solve the long-lasting enigma of the origin of MCs. Demonstration of MC derivation from bone marrow precursors could be established in 1977 when Yukihiko Kitamura's group first showed that, using the abnormal giant cellular granules of beige mice (C57BL-Bg^J/Bg^J) as a traceable marker, tissue MCs were found to develop from grafted beige bone marrow in irradiated wild-type recipient mice (Kitamura et al., 1977).

This discovery prompted further investigations on the origin of MC lineage leading to the concept that MCs were tissue-homing leukocytes. The paradigm was developed that they arise from pluripotent hematopoietic stem cells, circulate in the blood as agranular progenitors, and then acquire their mature phenotype within tissues.

The main growth factor governing tissue MC development and complete differentiation was shown to be the KIT receptor ligand or stem cell factor (SCF). The importance of SCF as a MC growth factor was underlined by the fact that mice with certain loss-of-function mutations affecting either SCF or its receptor KIT were devoid of MCs. Indeed, lack of expression of a functional KIT receptor due to spontaneous mutation in both copies of *Kit*, as it occurred in genetically mast-deficient WBB6F1-Kit^{W}-$Kit^{W\text{-}v}$ mice (W/W^v mice), resulted in a virtual absence of tissue MCs (Kitamura et al., 1978).

This important finding stimulated further studies on the genetics of the KIT-SCF system. These studies provided a series of MC-lacking mouse strains which revealed extremely useful to study different aspects of MC function. Indeed, lack of MCs in *Kit*-mutant mice could be selectively repaired by the adoptive transfer of genetically compatible wild-type or mutant MCs derived from in vitro cultures to create the so called MC "knock-in" mice (Nakano et al., 1985).

Most of our current knowledge on MCs is actually derived from MC "knock-in" mouse models which have allowed researchers for testing and verifying whether MCs contribute to specific functions. By the way, the finding that basophils lacked the KIT receptor and were unaffected by the SCF shaped the concept that the developmental pathways of MCs and basophils were different.

In the 1980s, many investigations focused on the composition of mst cell granules and the capacity of stimulated MCs to release cytokines, chemokines and growth factors. The concept of MC heterogeneity was further defined at this stage of MC research (Bienenstock et al., 1983).

Like rodent MCs, human MCs were also found to express morphological, biochemical and functional heterogeneity. The first evidence that MCs contained proteases was provided by Gomori (1953). He developed enzyme histochemical techniques for detecting esterase activity inside of cells in sections of fixed tissues and was able to recognize that MCs stained intensely with such procedures. By 1960, two proteases with chymotrypsin- and trypsin-like activity were identified in MCs (Benditt, 1956; Benditt and Arase 1959; Glenner and Cohen 1960). Enzyme activity was recognized to localize within intracellular granules.

These enzymes were purified in the 1980s and renamed tryptase and chymase (Schwartz et al., 1981; Schechter et al., 1986). It soon appeared that MCs from different anatomical sites contained different profiles of these enzymes as well as of other proteases identified in the meantime. Human MCs were thus divided into two subtypes depending on the expression of different proteases in their granules (and other functional features) (Irani et al., 1986).

MCs, which contained tryptase only, were designated as MCs_T or "immune cell-associated" MCs. They were predominantly located in the respiratory and intestinal mucosa, where they colocalized around T lymphocytes. MCs that contained both tryptase and chymase, along with other proteases such as carboxypeptidase A and cathepsin G, were referred to as MCs_{TC}. They were predominantly found in connective tissue areas, such as skin, submucosa of stomach and intestine, breast parenchyma, myocardium, lymph nodes, conjunctiva, and synovium.

These two subsets of human MCs differed also in terms of their mediator content and reactivity. A third type of MC, called MC_C was also identified. This MC expressed chymase without tryptase and resided mainly in the submucosa and mucosa of the stomach, small intestinal submucosa and colonic mucosa (Irani and Schwartz, 1994). Interestingly, human MCs_T were seen to correspond most closely to rodent MMCs, whereas MCs_{TC} resembled rodent CTMCs. It was later recognized that the concept of MC heterogeneity was not limited to staining properties but also involved functional characteristics.

Beginning from the end of the 1980s, it progressively emerged that granules in MCs contained a series of highly active biological compounds, such as cytokines, chemokines, and growth factors.

In 1989, a series of groups investigating on MC responses to various activators reported that stimulated MCs produced and released interleukin (IL)-3, IL-4, IL-5, IL-6 and granulocyte/macrophage-colony stimulating factor (GM-CSF) (Burd et al., 1989; Plaut et al., 1989; Wodnar-Filipowicz et al., 1989), and that, this could occur in the absence of MC degranulation.

Shortly thereafter, Gordon and Galli (1991) reported that MCs were a biologically relevant source of both preformed and antigen-induced tumor necrosis factor (TNF)-α. These discoveries were central to set MCs in the midpoint of a complex series of inflammatory and immunological pathways associated to host defensive responses, as revealed by two seminal papers appearing in 1996 which demonstrated that MCs were essential to survival in a mouse model of sepsis (Malaviya et al., 1996; Echternacher et al., 1966).

On the basis of these and other succeeding findings, it soon appeared that MCs were endowed with a large series of preformed and newly synthesized mediators capable to exert different immunological and nonimmunological functions.

REFERENCES

Askenase, P.W., 1977. Immune inflammatory responses to parasites: the role of basophils, mast cells and vasoactive amines. Am. J. Trop. Med. Hyg. 26, 96–103.

Beaven, M.A., 2009. Our perception of the mast cell from Paul Ehrlich to now. Eur. J. Immunol. 39, 11–25.

Behrens, M., Taubert, M., 1952. Der Nachweis von Heparin in den basophilen Leukocyten. Klin Wochenschr 30, 76–78.

Benditt, E.P., Wong, A.L., Arase, M., et al., 1955. 5-hydroxytryptamine in mast cells. Proc. Soc. Exp. Biol., N.Y. 90, 303–304.

Benditt, E.P., 1956. An enzyme in mast cells with some properties resembling chymotrypsin. Fed. Proc. 15, 507.

Benditt, E.P., Arase, M., 1959. An enzyme in mast cells with properties like chymotrypsin. J. Exp. Med. 110, 451–460.

Bienenstock, J., Befus, A.D., Denburg, J., et al., 1983. Mast cell heterogeneity. Monogr. Allergy 18, 124–128.

Blank, U., Ra, C., Miller, L., et al., 1989. Complete structure and expression in transfected cells of high affinity IgE receptor. Nature 337, 187–189.

Brocklehurst, W.E., 1960. The release of histamine and the formation of slow reacting substance (SRS-A) during anaphylactic shock. J. Physiol 151, 416–435.

Burd, P.R., Rogers, H.W., Gordon, J.R., et al., 1989. Interleukin 3-dependent and -independent mast cells stimulated with IgE and antigen express multiple cytokines. J. Exp. Med. 170, 245–257.

Cass, R., Riley, J.F., West, G.B., et al., 1954. Heparin and histamine in mast cell tumours from dogs. Nature, Lond. 174, 318.

Crivellato, E., Finato, N., Ribatti, D., et al., 2005. Do mast cells affect villous architecture? Facts and conjectures. Histol. Histopathol. 20, 1285–1293.

Crivellato, E., Finato, N., Isola, M., et al., 2006. Number of pericryptal fibroblasts correlates with density of distinct mast cell phenotypes in the crypt lamina propria of human duodenum: implications for the homeostasis of villous architecture. Anat. Rec., A 288, 593–600.

Day, M., Green, J.P., 1962. The uptake of biogenic amines by neoplastic mast cells in culture. J. Physiol. 164, 227–237.

Dvorak, H.F., Dvorak, A.M., 1975. Basophilic leukocytes: structure, function and role in disease. In: Lichtman, M.A. (Ed.), Clinics in Haematology, IV. Saunders WB Co, London, p. 651.

Dvorak, A.M., Kissel, S., 1991. Granule changes of human skin mast cells characteristic of piecemeal degranulation and associated with recovery during wound healing in situ. J. Leuk. Biol. 49, 197–210.

Echternacher, B., Mannel, D.N., Hultner, L., 1966. Critical protective role of mast cells in a model of acute septic peritonitis. Nature 381, 75–77.

Ehrlich P. Beitrage zur Theorie und Praxis der histologiscen Farbung. Thesis. Leipzig, Leipzig University, 1878.

Ehrlich, P., 1879. Beitrage zur Kenntnis der granulierten Bindgewebenzellen und der eosinophilen Leukocyten. Archiv. Anat. Physiol. 3, 166–169.

Ehrlich P, Lazarus A., 1898. Die Anemie, 1. Normale und Patologische Histologie des Blutes. Holder, Wien (revised and republished 1909).

Enerbäck, L., 1966a. Mast cells in rat gastrointestinal mucosa. 1. Effects of fixation. Acta Pathol. Microbiol. Scand. 66, 289–302.

Enerback, L., 1966b. Mast cells in rat gastrointestinal mucosa. 2. Dyebinding and metachromatic properties. Acta Pathologica. Microbiol. Scand. 66, 303–312.

Enerback, L., 1986. Mast cell heterogeneity: the evolution of the concept of a specific mucosal mast cell. In: Befus, A.D., Bienenstock, J., Denburg, J.A. (Eds.), Mast Cell Differentiation and Heterogeneity. Raven Press, New York, pp. 1–26.

Erspamer, V., Asero, B., 1952. Identification of enteramine, specific hormone of enterochromaffin cells, as 5-hydroxytryptamine. Nature 169, 800–801.

Fawcett, D.W., 1954. Cytological and pharmacological observations on the release of histamine by mast cells. J. Exp. Med. 100, 217–224.

Feldberg, W., Kellaway, D.H., 1983. Liberation of histamine and formation of lyscithin-like substances by cobra venom. J. Physiol. 94, 187–226.

Furano, A.V., Green, J.P., 1964. The uptake of biogenic amines by mast cells of the rat. J. Physiol. 170, 263–271.

Galli, S.J., 1990. Biology of disease. New insights into "the riddle of mast cells": microenvironmental regulation of mast cell development and phenotypic heterogeneity. Lab. Invest. 62, 5–33.

Galli, S.J., Kalesnikoff, J., Grimbaldeston, M.A., et al., 2005. Mast cells as "tunable" effector and immunoregulatory cells: recent advances. Ann. Rev. Immunol. 23, 749–786.

Glenner, G.G., Cohen, L.A., 1960. Histochemical demonstration of a species-specific trypsin-like enzyme in mast cells. Nature 185, 846–847.

Gomori, G., 1953. Chloroacyl esters as histochemical substrates. J. Histochem. Cytochem. 1, 469–470.

Gordon, J.R., Galli, S.J., 1991. Release of both preformed and newly synthesized tumor necrosis factor alpha (TNF-α)/cachectin by mouse mast cells stimulated via the FcεRI. A mechanism for the sustained action of mast cell-derived TNF-α during IgE-dependent biological responses. J. Exp. Med. 174, 103–107.

Graham, H.T., Lowry, O.H., Wheelwright, F., et al., 1955. Distribution of histamine among leukocytes and platelets. Blood 10, 467–481.

Hardy, W.B., Wesbrook, F.F., 1895. The wandering cells of the alimentary canal. J. Physiol. 18, 490–524.

Holmgren, H., Wilander, O., 1937. Beitrag zur Kenntniss der Chemie und Funktion der Ehrlichschen Mastzellen. Z Mikrosk Anat Forsch 42, 242–278.

Irani, A.M., Schwartz, L.B., 1994. Human mast cell heterogeneity. Allergy Proc. 15, 303–308.

Irani, A.M., Schechter, N.M., Craig, S.S., et al., 1986. Two types of human mast cells that have distinct neutral protease composition. Proc. Natl. Acad. Sci. USA. 83, 4464–4468.

Ishizaka, K., Ishizaka, T., 1967. Identification of γE-antibodies as a carrier of reaginic reactivity. J. Immunol. 99, 1187–1198.

Ishizaka, T., Ishizaka, K., Orange, R.P., et al., 1970. The capacity of human immunoglobulin E to mediate the release of histamine and slow reacting substance of anaphylaxis (SRS-A) from monkey lung. J. Immunol. 104, 335–343.

Jolly, M.J., 1900. Clasmatocytes et mastzellen. Compte Rendus Société de Biologie (Paris) 52, 437–455.

Jorpes, E., Holmgren, H., Wilander, O., 1937. Über das Vorkommen von Heparin in den Gefässwänden und in den Augen. Ein Beitrag zur Physiologie der Ehrlichschen Mastzellen. Z Mikrosk Anat Forsch 42, 279–301.

Kirschenbaum, A.S., Kessel, S.W., Goff, J.P., et al., 1991. Demonstration of the origin of human mast cells from CD34+ bone marrow progenitor cells. J. Immunol. 146, 1410–1415.

Kitamura, Y., Shimada, M., Hatanaka, K., et al., 1977. Development of mast cells from grafted bone marrow cells in irradiated mice. Nature 268, 442–443.

Kitamura, Y., Go, S., Hatanaka, K.U., 1978. Decrease of mast cells in W/Wv mice and their increase by bone marrow transplantation. Blood 52, 447–452.

Lewis, R.A., Soter, N.A., Diamond, P.T., et al., 1982. Prostaglandin D2 generation after activation of rat and human mast cells with anti-IgE. J. Immunol. 129, 1627–1631.

Liu, J., Divoux, A., Sun, J., et al., 2009. Genetic deficiency and pharmacological stabilization of mast cells reduce diet-induced obesity and diabetics in mice. Nat. Med. 15, 940–945.

Malaviya, R., Ikeda, T., Ross, E., et al., 1996. Mast cell modulation of neurtophil influx and bacterial clearance at sites of infection through TNF-α. Nature 381, 77–80.

Michels, N.A., 1938. The mast cells. In: Downey, H. (Ed.), Handbook of Hematology, 1. Hoeber, New York, pp. 232–354.

Mota, I., Vugman, I., 1956. Effects of anaphylactic shock and compound 48/80 on the mast cells of the guinea pig lung. Nature 177, 427–429.

Nakano, T., Sonoda, T., Hayashi, C., et al., 1985. Fate of bone marrow-derived cultured mast cells after intracutaneous, intraperitoneal, and intravenous transfer into genetically mast cell-deficient W/Wv mice: evidence that cultured mast cells can give rise to both connective tissue type and mucosal mast cells. J. Exp. Med. 162, 1025–1043.

Orange, R.P., Austen, W.G., Austen, K.F., 1971. Immunological release of histamine and slow-reacting substance of anaphylaxis from human lung. I. Modulation by agents influencing cellular levels of cyclic 3′,5′-adenosine monophasphate. J. Exp. Med. 134, 136s–148ss.

Parratt, J.R., West, G.B., 1957. 5-hydroxytryptamine and tissue mast cells. J. Physiol. 137, 169–178.

Plaut, M., Pierce, J.H., Watson, C.J., et al., 1989. Mast cell lines produce lymphokines in response to cross-linkage of FcεRI or to calcium ionophores. Nature 339, 64–67.

Prausnitz, C., Küstner, H., 1921. Studien über die Überempfindlichkeit. Zentralbl Bakteriol Parasitenk Infektionsk Abt 1 86, 160–169.

Rapport, M., Green, A., Page, I., 1948. Crystalline serotonin. Science 108, 329–330.

Riley, J.F., West, G.B., 1953. The presence of histamine in tissue mast cells. J. Physiol. 120, 528–537.

Riley, J.F., 1954. Heparin, histamine and mast cells. Blood 9, 1123–1126.

Riley, J.F., 1955. Pharmacology and functions of mast cells. Pharmacol. Rev. 7, 267–277.

Riley, J.F., 1963. Functional significance of histamine and heparin in tissue mast cells. Ann. N.Y. Acad. Sci. 103, 151–163.

Roberts, L.J., Lewis, R.A., Oates, J.A., et al., 1979. Prostaglandin tromboxane, and 12-hydroxy-5,8,10,14-eicosatetraenoic acid production by ionophore-stimulated rat serosal mast cells. Biochim. Biophys. Acta 575, 185–192.

Rocha, E., Silva, M., 1942. Concerning the mechanism of anaphylaxis and allergy. Br. Med. J. 1, 779–784.

Schechter, N.M., Choi, J.K., Slavin, D.A., et al., 1986. Identification of a chymotrypsin-like proteinase in human mast cells. J. Immunol. 137, 962–970.

Schwartz, L.B., Lewis, R.A., Austen, K.F., 1981. Tryptase from human pulmonary mast cells. Purification and characterization. J. Biol. Chem. 256, 11939–11943.

Westphal, E., 1891. Uber mastzellen. In: Rhrlich, P. (Ed.), Farbenanalytische Untersuchungen. Hirschwald, Berlin, pp. 17–41.

Wodnar-Filipowicz, A., Heusser, C.H., Moroni, C., 1989. Production of the haematopoietic growth factor GM-CSF and interleukin-3 by mast cells in response to IgE receptor-mediated activation. Nature 339, 150–152.

FURTHER READING

Bienenstock, J., 1988. An update on mast cell heterogeneity. J. Allergy Clin. Immunol. 81, 763–769.

Dvorak, A.M., 2005. Ultrastructural studies of human basophils and mast cells. J. Histochem. Cytochem. 2005 (53), 1043–1070.

Feger, F., Varadaradjalou, S., Gao, Z., et al., 2002. The role of mast cells in host defense and their subversion by bacterial pathogens. Trends Immunol. 23, 151–158.

Fureder, W., Agis, H., Willheim, M., et al., 1995. Differential expression of complement receptors on human basophils and mast cells. Evidence for mast cell heterogeneity and CD88/C5aR expression on skin mast cells. J. Immunol. 155, 3152–3160.

Nakae, S., Suto, H., Hakurai, M., et al., 2005. Mast cells enhance T cell activation: importance of mast cell-derived TNF. Proc. Natl. Acad. Sci. USA. 102, 6467–6472.

Rodewald, H.R., Dessing, M., Dvorak, A.M., et al., 1996. Identification of a committed precursor for the mast cell lineage. Science 271, 818–822.

Shanahan, F., Denburg, J.A., Fox, J., et al., 1985. Mast cell heterogeneity: effect of neuroenteric peptides on histamine release. J. Immunol. 135, 1331–1337.

Sher, A., Hein, A., Moser, G., et al., 1979. Complements receptors promote the phagocytosis of bacteria by rat peritoneal mast cells. Lab. Invest. 41, 490–499.

Talkington, J., Nickell, S.P., 2001. Role of Fgc receptors in triggering host-cell activation and cytokine release by Borrelia burgdorferi. Infect. Immun. 69, 413–419.

CHAPTER 10

The Origins of Lymphatic Vessels: An Historical Review

10.1 LYMPHANGIOGENESIS

The term "lymphangiogenesis" describes any growth-inducing events of lymphatics, such as proliferation and tube formation on plastic (Leak and Jones, 1994), and invasion into collagen gels (Tan et al., 1998). The lymphatic system includes a wide network of capillaries, collecting vessels and ducts that permeate most of the organs (Ryan et al., 1986). Unlike the blood vasculature, which forms a continuous loop, the lymphatic system is an open-ended, one-way transit system. It assists in maintaining the blood volume, carries cells, interstitial fluid components, and metabolites that leak from capillaries and return them to the venous circulation via the thoracic duct.

Furthermore, it entails part of the immune system by continuously circulating the white blood cells within the lymphoid organs and bone marrow and transporting antigen-presenting cells. Endothelial receptors and binding proteins are involved in this trafficking. Like developing blood vessels, the first lymphatics consist only of endothelial cells. The origin of embryo lymphatic endothelial cells is not known yet.

10.2 FIRST DESCRIPTION OF THE LYMPHATIC VESSELS

The first anatomical description of the lymphatic vessels as "milky veins" in the mesentery of a "well-fed" dog was published by Aselli (1627) (Fig. 10.1), professor of anatomy and surgery in Pavia at the same time that William Harvey described the blood circulation (Ribatti, 2009). In 1651, Jean Pecquet in Dieppe discovered the thoracic duct and its entry into the left subclavian vein, which for the first time described the correct route of the lymphatic fluid entry into blood circulation. At the same time, Olaus Rudbeck in Uppsala made similar observation in humans. He also discovered that lymphatic vessels are

Milestones in Immunology. DOI: http://dx.doi.org/10.1016/B978-0-12-811313-4.00010-3

Figure 10.1 A portrait of the Italian anatomist Gaspare Aselli.

distributed throughout the body and form a distinct vascular network. His observations, together with those made independently by Thomas Bartholin in Copenhagen, led to the first full description of the human lymphatic system, in which the role of lymphatic vessels in transporting fluid filtered from the blood was suggested. William Hunter (1718–83) believed, "The lymphatic vessels are the absorbing vessels, all over the body." William Hewson (1739–74), a pupil of Hunter, made extensive dissections of the lymphatic system in fishes, birds, and mammals, for which he was awarded the Copley Medal by the Royal Society in 1769 (Doyle, 2006). Hewson noted that lymph glands were absent in fishes, few in number in birds, and well developed only in mammals. Monro Secundus (1733–1817) was a Scottish anatomist who spent a short time working under Hunter in London and claimed in a publication entitled De venis lymphaticis valvulosis (1757) that he had discovered the anatomy and role of the lymphatic vessels.

10.3 FIRST SUCCESSFUL CULTURES OF LYMPHATIC ENDOTHELIUM AND LYMPHANGIOGENESIS IN VITRO

In 1984, pure cultures of lymphatic endothelium were first isolated from mesenteric lymphatic duct by Johnston and Walker (1984) and by

Bowman et al. (1984) from a patient with a massive cervicomediastinal lymphangioma, where tubule formation and associated cyst-like structures are spontaneously formed. In 1985, Gnepp and Chandler isolated lymphatic endothelium from canine and human thoracic duct. Nicosia (1987) used rat thoracic duct collagen-embedded explants to isolate both lymphatic (intimal extension) and more hematic (adventitial origin) endothelial cells. Leak and Jones (1994) demonstrated lymphangiogenesis in vitro in normal lymphatic cultures in a delayed postconfluent state or earlier after collagen I was added to the culture medium.

Identification of cell surface markers that allow easy distinction between lymphatic and blood vascular endothelium has led to the development of other techniques for the isolation of pure endothelial cell populations. Lymphatic endothelial cells have been isolated by positive selection using antibodies to podoplanin, vascular endothelial growth factor receptor-3 (VEGFR-3), or lymphatic vessel endothelial hyaluronan receptor-1 (LYVE-1), and by a negative selection with antibodies to CD34.

10.4 LYMPHATIC VESSELS DEVELOP FROM EMBRYONIC VEINS OR FROM MESENCHYMAL PRECURSOR CELLS

Based upon results obtained by India ink injection experiments in pig, Sabin (1902, 1904) proposed that isolated primitive lymph sacs originated from endothelial cells that bud from the cardinal vein during early development. To quote Sabin (1911), "Lymphatics are modified veins. They are vessels lined by an endothelium which is derived from the veins. They invade the body as do blood vessels and grow into certain constant areas; their invasion of the body is, however, not complete, for there are certain structures which never receive them. The lymphatic capillaries have the same relation tissue spaces as have blood capillaries. None of the cavities of the mesoderm, such as the peritoneal cavity, the various bursae and serous capillaries, forms any part of the lymphatic system. The lymphatic endothelium once formed is specific. Like blood vessels, the lymphatics are for the most part closed vessels."

Sabin showed that, starting from the region of each of the primitive sacs, there is in the skin a gradually increasing zone of injectable lymphatics which eventually covers the whole body (Fig. 10.2).

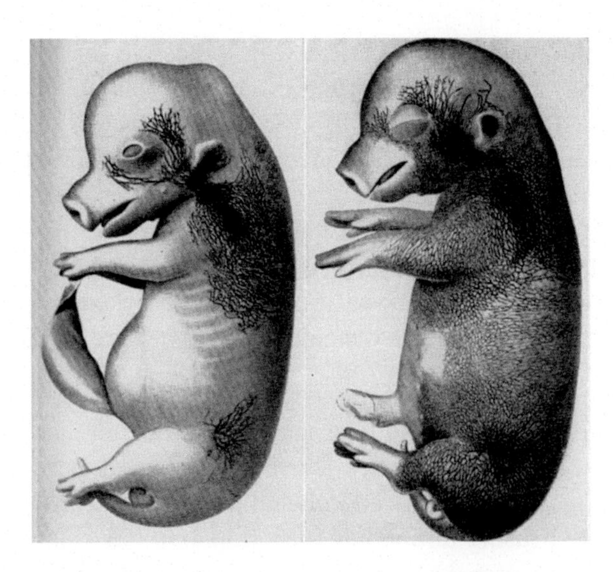

Figure 10.2 The growth of the lymphatic vessels of the embryo of the pig depicted by India ink injection. Starting from the region of each of the primitive sacs, the skin shows a gradually increasing zone of injectable lymphatics which eventually covers the whole body. Reproduced from Sabin, F.R., 1904. On the development of the superficial lymphatics in the skin of the pig. Am. J. Anat. 3, 183–195.

The India ink injection method did not show up any isolated lymphatics, which suggest that growth is toward the periphery by the sprouting of preexisting endothelium. Clark (1909a) observed that from the walls of the lymphatics line pointed projections extend at various intervals and of varying lengths, whereas the tip ends in one or more pointed processes. This living tip is always changing, and the nuclear thickening in the wall of the capillary is perpetually altering shape and position (Fig. 10.3).

Studies on mammalian embryos have shown that there are eight lymph sacs; three paired and two unpaired (Sabin, 1909). The paired ones are the jugular, subclavian, and posterior lymph sacs, and the unpaired are the *Cisterna chyli* and the retroperitoneal (mesenterial) lymph sac. In the human, the subclavian lymph sac is an extension of the jugular lymph sac (Sabin, 1909). Except for the *Cisterna chyli*, the lymph sacs develop into primary lymph nodes. The two jugular lymph sacs (JLS) were thought to develop at the junction of the subclavian and anterior cardinal veins. These observations were confirmed by Lewis in rabbits and humans (Lewis, 1905, 1909).

Lymph sacs are venous derivatives that grow by sprouting into all parts of the body, except for the central nervous system and the bone marrow, which remains free of lymphatics. According to this model,

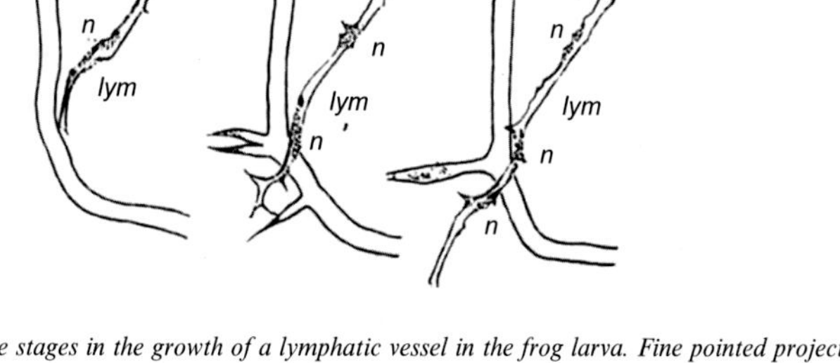

Figure 10.3 Successive stages in the growth of a lymphatic vessel in the frog larva. Fine pointed projections extend at various intervals and of various lengths from the walls of the lymphatics, whereas the tip ends in one or more pointed processes. b.v., *blood vessels;* lym, *lymphatics;* n, *nucleus.* Reproduced from Clark, E.R., 1909. Observation on living growing lymphatics in the tail of the frog larva. Anat. Rec. 3, 183–198.

the peripheral lymphatic system originates from the primary lymph sacs and spreads by endothelial sprouting into the surrounding tissues and organs, where local capillaries are formed (Sabin, 1902). Lymph sacs later will give rise to primary lymph nodes (Lewis, 1905).

Clark and Clark (1932, 1933, 1937, 1938) documented the extension of lymphatic capillaries by outgrowth from existing lymph vessels in rabbit ear transparent chambers, accordingly to the model developed by Sandison (1924). Using high magnification, they observed that the lymphatic capillary wall was of delicate endothelium in which the nuclear thickening stood out distinctly as clear lens-shaped structures, or as large rounded swellings which bulge into the lumen. In the course of hours, lymphatic capillary increased in length and a lumen extended into a new sprout. Abundant new lymphatic capillaries in the area under observation appeared some what later (approximately 30 days) than blood capillaries (5–10 days). This growth process can also apparently be reversed. In fact, many of the newly formed lymphatic capillaries may retract or disappear.

Pullinger and Florey (1937), after examining the proliferation of lymphatics in inflammation and abscesses in the living mouse ear, emphasized that unless opaque or colored materials were used to identify lymphatics, they might otherwise be overlooked. They also noted that lymphatics consistently connected to lymphatics, veins to veins, and arteries to arteries without intermingling. Bellman and Odén (1959) documented via contrast microlymphangiography the time course and extent of newly formed lymphatics in circumferential wounds of the rabbit ear.

An alternative model proposed that lymphatic vessels developed independently from mesenchymal precursor cells (but not from embryonic veins) and that connections with veins were only established later on during development (Huntington, 1908; McClure, 1908; Huntington and McClure, 1910; Kampmeier, 1912). This view depends largely upon the results of studies of serial sections with graphic wax-plate reconstructions and has been supported by Balankura (1951). Accordingly to this model, lymphatic vessels originate by fusion of mesenchymal spaces into a primitive lymphatic network, which spreads centripetally (centripetal theory) and then establishes a connection to the venous system.

10.5 EMBRYO LYMPHANGIOGENESIS

The development of the embryo lymphatic system has been studied descriptively, but experimental studies have not been performed until recently (Schneider et al., 1999; Wigle et al., 2002; Wilting et al., 2001). Studies on living animals through serial sectioning and injection methods have shown that early anlagen of the lymphatics of birds, mammals, and humans are the lymph sacs which develop in close association with the venous system (Ranvier, 1895; van der Putte, 1975). Due to their specific location, these are the jugular, subclavian, posterior and retroperitoneal (mesenterial) lymph sacs, and the *cisterna chyli*. The lymph sacs derive by fusion of lymphatic capillaries filled with stagnant blood, which has prompted the speculation that the lymph sacs derive from veins (Lewis, 1905; Miller, 1912).

However, different theories about the origin of the lymphatics have been proposed. Historically, the best accepted is the Sabin's theory

(1902, 1904). He suggested that early in fetal development, isolated primitive lymph sacs originate by endothelial cell budding from embryo veins. Thus, the peripheral lymphatic vessels spread from these primary lymph sacs by endothelial sprouting into surrounding tissues and organs where local lymphatic capillaries form. Accordingly, lymphatic endothelial cells are derived exclusively from the endothelium of the venous system. This view seems to be supported by Wigle et al. (2002).

In contrast, Huntington (1908), McClure (1908), and Kampmeier (1912) suggested that the primitive lymphatic vessels arise in the mesenchyme from putative lymphangioblasts (independent on the veins) by confluence of "lymphatic clefts," fuse with the lymph sacs by centripetal growth and secondarily established venous connections. This theory is in line with findings by Schneider et al. (1999), who have demonstrated through the quail-chick chimera system (Le Douarin, 1969) that lymphangioblasts are present in the avian wing bud before the emergence of the jugulo-axillary lymph sacs. They grafted paraxial mesoderm of 2-day-old quail embryo into the shoulder region of 3-day-old chick embryo and studied the integration of lymphangioblasts into the jugulo-axillary lymph sac. Then, the host embryo was reincubated until day 6.5 (grafting experiments). Due to morphological and molecular characteristics (QH1 staining, a specific marker of quail—but not of chick—endothelial cells) (Pardanaud et al., 1987), the jugular lymph sac was clearly distinguished from the jugular vein. In all specimens, quail endothelial cells integrating into the endothelial lining of chick vessels were observed. Specifically, two specimens showed quail cells being both in jugular vein and lymph sacs without possibility of estimating whether lymphangioblasts had directly integrated into the lymph sac, or whether cells had been first integrated into the jugular vein and then given rise to the jugular lymph sac. However, the serial sectioning of one specimen revealed that no quail cells had integrated into the jugular vein, and that these cells were in the jugular lymph sac (at a great distance from the jugular vein) indicating that they had been integrated directly into the lymph sacs. Therefore, the early lymph sacs may arise as sprouts from adjacent veins, but additionally, mesenchymal lymphangioblasts contribute to the growing lymph sacs.

To determine the origin of the lymphatic endothelium, the distal wing buds of 3.5-day-old chick embryo were grafted homotopically into 3–3.5-day-old quail embryo. The VEGFR-3 and QH1 double staining of the 10-old day chimeric wings revealed that lymphatics were formed by both chick and quail endothelial cells (Schneider et al., 1999).

Papoutsi et al. (2001) studied the lymphangiogenic power of the splanchnic mesoderm of the avian allantoic bud of 3-day-old quail embryo grafted onto the chick embryo chorioallantoic membrane (CAM). The chick CAM contained areas where the endothelium of blood vessels and lymphatics were of quail origin (QH1 and VEGFR-3 double positive). The lymphatics were located in their typical position around arteries and veins, demonstrating lymphangiogenic potential of the allantoic mesoderm long before the development of the posterior lymph sacs.

10.6 STRUCTURAL FEATURES OF LYMPHATIC SYSTEM

The initial lymphatic vessels (also called "absorbing lymphatic vessels" and, improperly, "lymphatic capillaries") are blind-ended structures that optimally suit fluid and particle uptake. Like blood capillaries, they are formed by a single nonfenestrated endothelial cell layer, but they differ in that: (1) initial lymphatic vessels usually possess a more irregular and wider lumen (10–60 μm in diameter); (2) their endothelium has an extremely attenuated cytoplasm, except in the perinuclear region (Fig. 10.4); (3) they are not encircled by pericytes and have

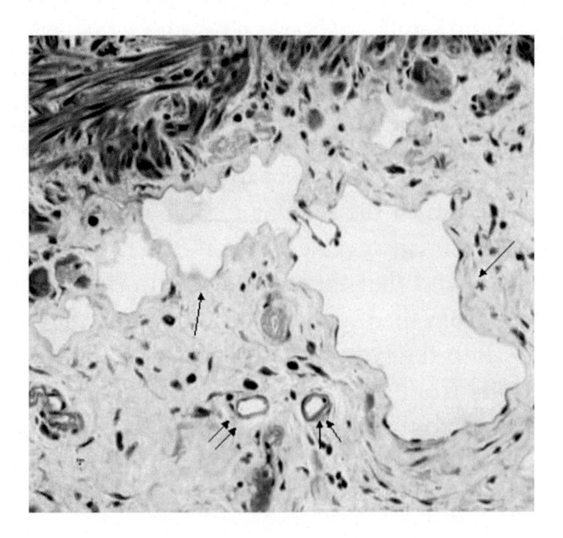

Figure 10.4 Initial lymphatic vessels (arrows) in the submucosa of human colon; the largest contains a well-formed valve. Surrounding blood microvessels (double arrows) are much smaller.

absent or poorly developed basal lamina; (4) their tight junctions and adherens junctions, which are the major intercellular junctions in blood vessels, are not as frequently seen; (5) these junctions different from those of blood vessels, which are typically involved in maintaining firm cell–cell adhesion to connect adjacent endothelial cells over entire cell boundaries, represent focal points of adhesion (Leak, 1970; Schmelz et al., 1994); (6) their intimate association with the adjacent interstitial areas. In effect, lymphatic endothelial cells are closely linked to surrounding connective tissue by fine (10–12 nm) "anchoring filaments" (Gerli et al., 1991) (Fig. 10.5). These filaments are attached to the cell abluminal surface and extended deeply into the connective tissue, firmly attaching endothelium to extracellular matrix fibers, which is thus highly sensitive to interstitial stresses. It has long been thought that an increase in the interstitial fluid volume causes the filaments to exert radial tension on the initial lymphatic vessel and pull open overlapping intercellular junctions (formed by extensive superimposing of adjacent endothelial cells), hence favoring interstitial drainage (Aukland et al., 1993). This mechanism creates a "tissue pump" or a slight and temporary pressure gradient that enables lymph formation (Ikomi et al., 1996).

Anchoring filaments are made of fibrillin (Solito et al., 1997; Gerli et al., 2000), a large glycoprotein that contains the tripeptide Arg-Gly-Asp (RGD) motif capable of binding $\alpha_v\beta_3$ integrins. The latter are transmembrane glycoproteins that cluster at focal adhesion plaques. Extracellular matrix stimuli transmitted by the cytoplasmic tail of integrins into the cell may trigger several signaling proteins including

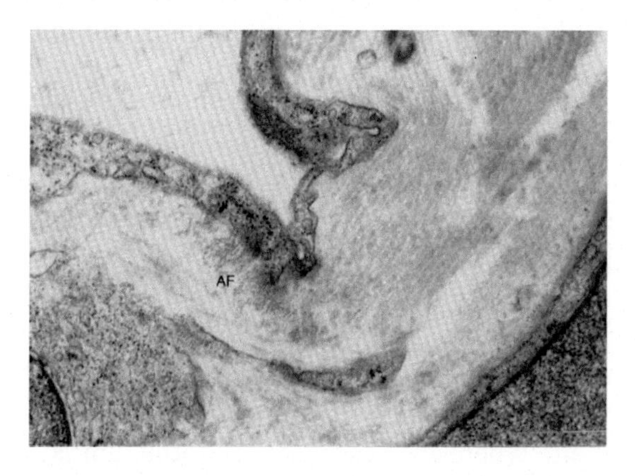

Figure 10.5 Anchoring filaments (AF) in a precollector of the human thigh.

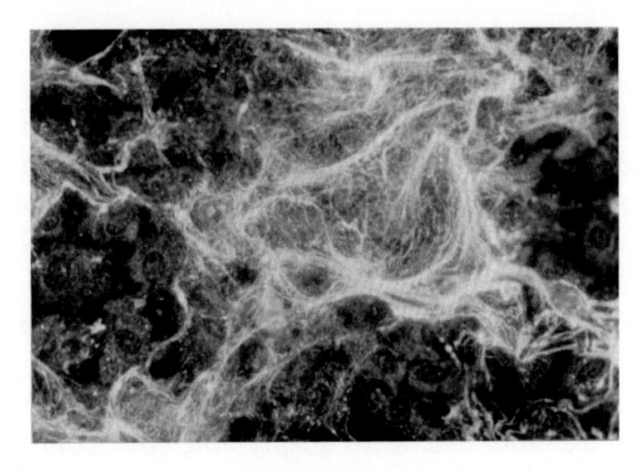

Figure 10.6 Immunofluorescence staining of cultured lymphatic endothelial cells deposit fibrillin, the major component of anchoring filaments, in the underlying extracellular matrix.

Focal Adhesion Kinase (FAK). Upon FAK activation, actin-mediated cytoskeletal rearrangements occur, and permeability is also probably affected. Thus, the lymphatic endothelium through fibrillin-containing anchoring filaments may play a much more complex role in lymph formation than is currently supposed (Weber et al., 2002). Finally, cultured lymphatic endothelial cells are able to produce an extensive network of fibrillin-containing microfibrils into the underlying matrix (Fig. 10.6).

In light of their importance in lymphatic function, the composition and architecture of extracellular matrix are likely to play a critical role in lymphangiogenesis. This agrees with the evidence that extensive and chronic degradation of the extracellular matrix renders lymphatics nonresponsive to the changes in the interstitium and therefore causes dysfunction (Negrini et al., 1996).

Initial lymphatic vessels merge into larger vessels called "precollectors" (Sacchi et al., 1997), which represent the initial drainage routes of lymph. Precollectors are structurally characterized by the alternance of areas with the same structural simplicity as initial lymphatic vessels and areas with a well-developed muscular coat (Fig. 10.7). The former probably has an absorbing function, whereas the latter may be involved in lymph propulsion. Interestingly, the ultrastructural features of these different portions do not differ, and anchoring filaments are present in both. Retrograde flow is prevented by thin but well-formed valves, the only site where the lymphatic endothelium is underlined by a continuous basement membrane.

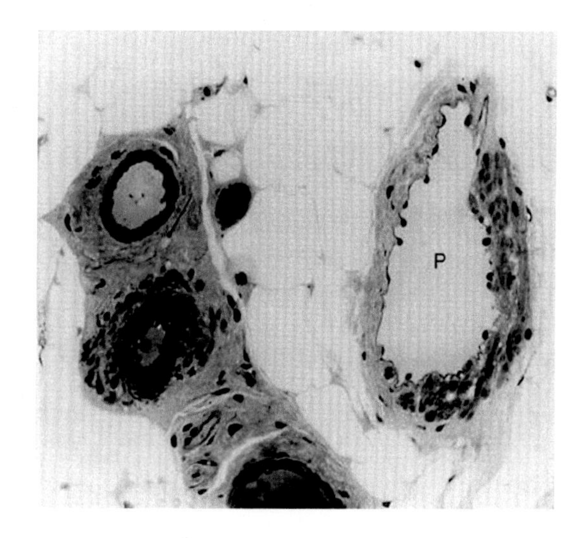

Figure 10.7 A precollector of the human thigh (P, on the right) and blood vessels (on the left). The precollector wall has no muscular coat on the left side.

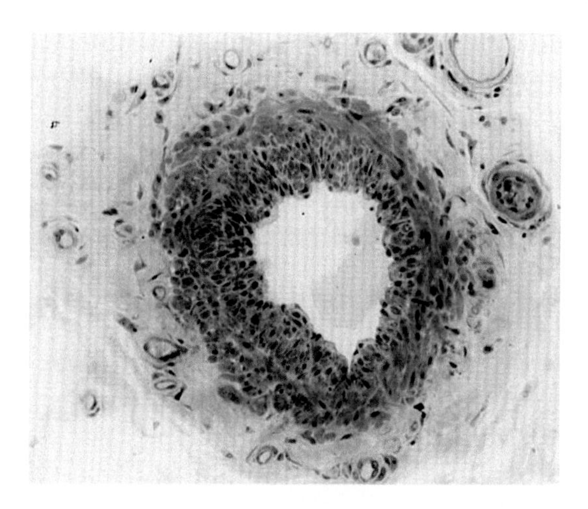

Figure 10.8 A collecting vessel of the human thigh surrounded by several blood microvessels. A continuous muscular coat enables distinction from precollectors. Original magnification: ×200.

Precollectors merge into collecting vessels (Fig. 10.8) which have a thick wall comprising a continuous layer of smooth muscle cells and may thus support a circumferential hoop stress. They also contain one-way valves that aid in lymph propulsion by preventing retrograde flow. Collecting vessels are interrupted by lymph nodes where lymph is filtered and can thus be distinguished in pre- and postnodal. They eventually drain, through larger trunks, into the thoracic duct and the right lymphatic duct that discharge lymph into the large veins at the base of the neck.

10.7 MARKERS OF LYMPHATIC ENDOTHELIAL CELLS

Molecular markers that unequivocally distinguish lymphatics from blood vessels are critical to further understand lymphatic vessel formation and function. Several methods have been suggested to discriminate blood and lymphatic vessels in histological sections. Of these, some are more reliable and well characterized than others.

The lymphatic endothelium has been characterized by using 5′-Nucleotidase (Nase)-alkaline phosphatase (ALPase) double staining (Kato, 1900). The lymphatic endothelium is marked by strong 5′-Nase activity (Fig. 10.9) that is significantly lower or absent in blood vessels, whereas this stain with ALPase.

Other techniques are toluidine blue staining following arterial perfusion-fixation and staining of basement membrane components, such as laminin and collagen type IV (Kubo et al., 1990). PAL-E stains not at all with the endothelium of large, medium-sized and small arteries, arterioles, and large veins, and does not stain the endothelial lining of lymphatic vessels and sinus histiocytes (Schlingemann et al., 1985).

Platelet endothelial cell adhesion molecule-1 (PECAM-1)/CD31 and von Willebrand factor (vWF) are molecules widely expressed in all endothelial cells (Sleeman et al., 2001). However, CD31/pathologische

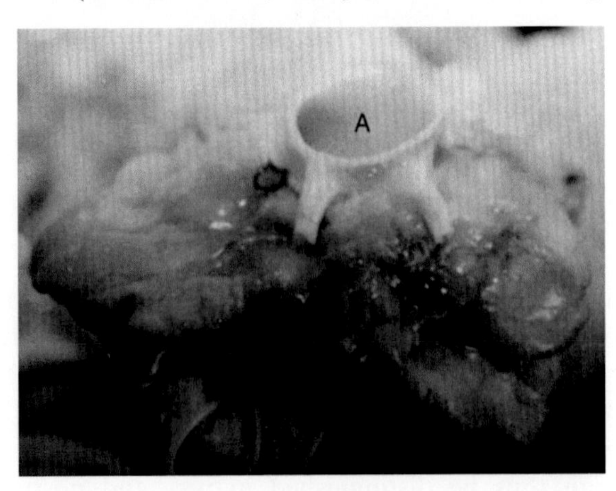

Figure 10.9 Caudal end of bovine thoracic duct stained with Evans blue in the periadventitial fat of the thoracic aorta (A).

anatomie Leiden-endothelium (PAL-E) double staining has been used for the detection of lymphatic endothelium (de Waal et al., 1997).

To sum up, so far none of the so-called lymphatic endothelium-specific markers has been shown to be absolutely specific for lymphatic vessels. The main proposed lymphatic markers are described below.

LYVE-1, a homolog of the CD44 hyaluronan (HA) receptor, is a new member of the LINK protein superfamily. It is a type I integral membrane polypeptide whose extracellular domain encodes a single cartilage Link module, the prototypic HA-binding domain conserved within all members of the Link or hyaladherin superfamily (Toole, 1990). The central core of the LYVE-1 Link module (C2–C3) is 57% identical to that of the human CD44 HA receptor, the only Link superfamily HA receptor described to date and the closest homolog of LYVE-1 (Banerji et al., 1999). These authors first described a restricted pattern of tissue expression by immunoperoxidase which revealed LYVE-1 lining of lymphatics in virtually every tissue. The greatest LYVE-1 expression was detected in submucosal lymph vessels underlying smooth muscle in the colon and in the lacteal vessels of intestinal villi (Banerji et al., 1999).

LYVE-1 is probably involved in the transport of HA across lymphatic endothelium, specifically in the movement of tissue HA from interstitium to lymph (Prevo et al., 2001). This is supported by the finding that LYVE-1 is present on both the luminal and abluminal sides of lymphatic capillaries, suggesting shuttling across the endothelium or transcytosis (Nicosia, 1987). Furthermore, LYVE-1 probably regulates the entry of leukocytes or tumor cells into the lumen of afferent lymphatic capillaries, promoting dissemination to regional lymph nodes (Jackson et al., 2001). The role of LYVE-1 in transcytosis of HA or in the regulation of leukocyte or tumor cell entry is largely hypothetical at the moment.

Mouta Carreira et al. (2001) have shown that LYVE-1 is not exclusive to the lymphatic vessels, being expressed by normal hepatic blood sinusoidal endothelial cells in mice and humans. LYVE-1 is not expressed by angiogenic blood vessels of human liver tumors and is weakly present on microvessels of regenerative hepatic nodules in cirrhosis, though both vessels are largely derived from the liver sinusoids. In addition, lymphatic vessels were found within the parenchymal fibrous areas that develop de novo in cirrhosis suggesting that cirrhosis is accompanied by lymphangiogenesis. Moreover, in human hepatocellular carcinoma and liver metastases, lymphatic vessels were restricted to the tumor margin and surrounding liver, whereas no lymphatics were observed in the tumor parenchyma or between tumor nodules.

Podoplanin (also known as OTS-8, T1 alpha, or E11 antigen) is a 43-kDa integral plasma membrane protein primarily found on the surface of rat glomerular epithelial cells (podocytes) and linked to flattening of foot processes that occurs in glomerular diseases (Matsui et al., 1999).

The biological function of podoplanin can be inferred from the knock-out studies (Schacht et al., 2003). Podoplanin null mice die at birth due to respiratory failure and have defects in lymphatic, but not blood vessel pattern formation. These defects are associated with diminished lymphatic transport, congenital lymphedema, and dilation of lymphatic vessels. These data identify podoplanin as a novel critical player that regulates different key aspect of lymphatic vasculature formation.

Breiteneder-Geleff et al. (1999) studied a panel of vascular tumors with antipodoplanin and anti-VEGFR-3 antibodies and showed overlapping patterns for these two markers in that: (1) affinity-purified rabbit antipodoplanin IgG specifically and exclusively immunolabeled endothelial cells of vessels that were clearly distinct from PAL-E-labeled blood vessels; (2) podoplanin-positive vessels were also selectively stained with anti-VEGFR-3 in double labeling experiments predominantly at the luminal surface; (3) dermal lymphatic capillaries identified by ultrastructural morphological criteria and immunoelectron microscopy were selectively labeled by antipodoplanin antibodies; (4) in benign lymphatic tumors, podoplanin localized in endothelial cells; (5) immunoblotting of lymphangiomas with antipodoplanin IgG revealed a ~38-kDa protein that was identical to podoplanin in human lung and isolated glomeruli.

Prospero homeobox protein-1 (Prox-1) is a marker for the subpopulation of endothelial cells that sprout to give rise to the lymphatic sacs during development (Wigle et al., 2002). Targeted deletion of the Prox-1 gene does not affect development of the vascular system, but the sprouting of the developing lymphatics is specifically ablated. Wigle et al. (2002) suggested that blood endothelial cell differentiation is independent on Prox-1 and that it is mandatory for establishment of lymphatic endothelial cell identity. In contrast to the wild-type embryo, in Prox-1 null embryo, the endothelial cells sprouting from the cardinal vein did not express the lymphatic vascular markers VEGFR-3, LYVE-1, or SLC chemokine. Instead, the mutant cells expressed the blood vascular markers laminin and CD34, indicating that these cells were not committed to the lymphatic endothelial cell lineage in the absence of Prox-1. Overall results suggest that Prox-1 activity is required for both maintenance of the sprouting of the venous endothelial cells and differentiation toward the lymphatic phenotype. Accordingly, it has been proposed that a blood vascular phenotype is the default fate of sprouting embryo venous endothelial cells; upon expression of Prox-1, these sprouting cells adopt a lymphatic vasculature phenotype.

Hong et al. (2002) identified Prox-1 as a master switch in the program specifying lymphatic endothelial cell fate, showing that Prox-1 expression upregulated the lymphatic markers podoplanin and VEGFR-3. Conversely, genes such as laminin, VEGF-C, neuropilin-1 (NRP-1), and intercellular adhesion molecule-1, whose expression parallel the blood endothelial cell phenotype, were downregulated.

β-chemokine receptor D6 is a human chemokine receptor which binds with high affinity to a wide array of proinflammatory β-chemokines, including RANTES, monocyte chemoattractant protein-1 and -3 (MCP-1 and MCP-3), but not human macrophage inflammatory protein-1 α (MIP-1α) (Nibbs et al., 1997). D6 was found on endothelial cells lining the lymphatic system in human dermis (Hub and Rot, 1998). D6 immunoreactivity was shown in the mucosa and other wall layers of the gut, in afferent lymphatics and subcapsular and medullary sinuses of lymph nodes, but not in blood vessel endothelial cells. D6 was also revealed in some malignant, highly vascularized tumors, suggesting their origin from lymphatic endothelial cells (Nibbs et al., 2001).

The pattern of D6 expression suggests a role for this molecule intervening in the regulation of chemokine-driven trafficking of leukocytes across lymphatics, or development and growth of lymphatics themselves. Moreover, the expression of this receptor on only a subset of lymphatics suggests a functional heterogeneity within the lymphatic vasculature. However, its role in the migration of tumor cells to regional lymph nodes needs further investigation.

Desmoplakin is a protein of the junction system connecting the very flat endothelial cells of lymphatic vessels (Schmelz et al., 1994). It is a marker for small lymphatic vessels (Sawa et al., 1999), being not expressed by larger lymphatic collecting ducts including the thoracic duct (Schmelz et al., 1994). However, desmoplakin can be detected in the junctions between cultered blood vessel endothelial cells (Kowalczyk et al., 1998). It thus remains to be determined whether desmoplakin is exclusively expressed in lymphatic capillaries, even if the lymphatic specificity of this marker has been established (Ebata et al., 2001).

Macrophage mannose receptor is mainly expressed on cells of the macrophage lineage where it mediates the uptake of microorganisms, host-derived glycoproteins and, viral endocytosis (Reading et al., 2000). The receptor is also expressed by mouse lymphatic endothelia whereas, by contrast, the expression in human lymphatic vessels needs to be further defined (Linehan et al., 2001). However, the biological role of this receptor in lymphatic vessels is unknown: perhaps, it plays a role in antigen capture and clearing in inflammation and immune-mediated processes (Groger et al., 2000).

Some of the putative highly lymphatic-specific markers have not as yet been confirmed. In this context, definitive identification of lymphatics should combine characteristic morphological features with consistent immunohistochemical profiles and leave open the question that, particularly in pathological states, lymphatics and blood vessels in tissue sections may become indistinguishable and need to be visualized as part of a functioning, communicating vascular network.

10.8 LYMPHANGIOGENESIS IN VIVO

Detailed descriptive studies revealed the mechanisms of lymphangiogenesis in different animal and human tissues. Clark and Clark (1932) documented the extension of lymphatic capillaries by outgrowth from existing lymph vessels in rabbit ear transparent chambers. Bellman and Odén (1959) documented by contrast microlymphangiography the time course and extent of newly formed lymphatics in circumferential wounds of the rabbit ear, including lymphatic bridging through newly formed scar (Odén, 1960). Subsequent studies documented restoration of distinctive ultrastructural features in newly regenerated lymphatic vessels, including the characteristic overlapping junctional complexes and Weibel–Palade bodies (storage depots for von vWF) (Magari and

Asano, 1978; Magari, 1987). Lymphatics are less labile than blood capillaries; they send out fewer sprouts, anastomose less frequently, and show much less tendency to retract or change in size or shape. Although temporary lymphedema occurs after lymphatic disruption, it usually resolves due to spontaneous regeneration or reconnection of lymphatics (Reichert, 1926; Slavin et al., 1999).

10.9 EXPERIMENTAL MODELS OF LYMPHANGIOGENESIS

10.9.1 Dye Injection and Mouse Tail Model

Fluorescence microlymphography technique enables the visualization of the initial lymphatic network in human dermis in vivo (Stanton et al., 1997). This technique was used to stain the lymph capillaries in the superficial layer of the skin of the mouse tail and with the use of densitometric image analysis, flow velocity along the tail was measured (Leu et al., 1994). Boardman and Swartz (2003) used the same experimental model to investigate the hypothesis that interstitial flow may initiate matrix remodeling and cell organization. They evaluated lymphatic function (fluid transport) and fluid channel architecture in situ with fluorescence microlymphangiography and observed the formation of functional channels in 2–3 weeks within an initially acellular gel. Immunohistochemical staining confirmed that these channels were lined with lymphatic endothelial cells.

10.9.2 Wound Healing

During wound healing lymphatic capillaries grow by sprouting from existing lymphatics, much in the same way as new blood capillaries sprout from existing capillaries or postcapillary veins during angiogenesis. The appearance of new lymphatic capillaries is always secondary to that of blood capillaries, although linear growth occurs at comparable speed (Clark, 1922; Clark and Clark, 1932). Paavonen et al. (2000) studied VEGFR-3 expression in experimental wounds made in dorsal skin of pigs. VEGFR-3-positive vessels were observed in the granulation tissue from day 5 onward, and very few VEGFR-3-positive lymphatic vessels persisted on day 9 and none on day 14. These results suggest that transient lymphangiogenesis occurs parallel with angiogenesis in healing wounds. Witmer et al. (1991) demonstrated that in granulation tissue, VEGFR-3 staining was observed in the proliferative superficial zone in plump blood vessel sprouts, in the intermediate zone in blood vessels and long lymphatic sprouts, and in deeper fibrous zone in large lymphatics, in a pattern demonstrating that lymphangiogenesis follows behing blood vessel angiogenesis.

10.9.3 The Chick CAM

The chick CAM is an extraembryonic membrane which serves as a gas-exchange surface, and its function is supported by a dense capillary network. Because of its extensive vascularization, the CAM has been broadly used to investigate the action of angiogenic and antiangiogenic molecules (Ribatti et al., 2001). Moreover, it contains a dense network of lymphatics accompanying the arteries and veins. Oh et al. (1997) studied the morphology and pattern of the lymphatics of normal CAM using semi- and ultrathin sectioning, immunohistochemical staining with anti-α-smooth muscle actin and fibronectin antibodies, in situ hybridization with VEGFR-2 and VEGFR-3 probes, and Mercox injection. They described a regular pattern of lymphatic vessels along all arteries, arterioles, and veins and observed an increasing number of lymphatic capillaries interconnecting larger lymphatic vessels in proportion with increasing size of the arteries. The lymphatics of CAM showed characteristics already described for normal lymphatics such as huge diameter, porous and very thin endothelial lining, and absence of poor basal lamina. Proliferation studies revealed a great amount of BrdU-labeled nuclei of lymphatic endothelial cells after 1 and 2 days of VEGF-C application. Finally, VEGFR-3 mRNA expression was shown to be restricted to the lymphatic endothelial cells of the differentiated CAM.

10.9.4 Transgenic Mouse Models

Overexpression of VEGF-C in the skin of transgenic mice resulted in lymphatic, but not vascular endothelial proliferation and vessel enlargement (Jeltsch et al., 1997). Transgenic mice overexpressing VEGF-C under the control of the rat insulin promoter (Rip) developed extensive lymphatic vessels around the Langerhans islets. When crossed with Rip1Tag2 transgenic mice that normally develop benign β-cell tumors, enlarged peritumoral lymphatics, and high rate of metastasis to pancreatic lymph nodes were observed (Mandriota et al., 2001). These findings demonstrate that VEGF-C induced lymphangiogenesis mediate tumor cell dissemination and the formation of lymph node metastases. Veikkola et al. (2001) created transgenic mice overexpressing a VEGFR-3-specific mutant of VEGF-C (VEGF-C156S) or

VEGF-D in epidermal keratinocytes under the keratin 14 promoter. Both transgens induced the growth of lymphatic vessels in skin, whereas the blood vessel architecture was not affected. These results indicate that stimulation of the VEGFR-3 signal tranduction pathway is sufficient to indice specifically lymphangiogenesis in vivo. Makinen et al. (2001a) showed that a soluble form of VEGFR-3 is a potent inhibitor of VEGF-c/VEGF-D signaling, and when expressed in the skin of transgenic mice, it inhibits fetal lymphangiogenesis and induces a regression of already formed lymphatic vessels, though the blood vasculature remains normal. Karkkainen et al. (2001) described lymphedema (Chy) mice with an inactivating VEGFR-3 mutation in their germ line, and swelling of the limbs because of hypoplastic cutaneous, but not visceral, lymphatic vessels. By using virus-mediated VEGF-C gene therapy, they were able to generate functional lymphatic vessels in the Chy mice, suggesting that growth factor gene therapy is applicable to human lymphedema. MBA-MD-435 breast cancer cells overexpressing VEGF-C were injected into nude mice and developed tumors with enlarged peritumoral lymphatics and high number of intratumoral lymphatics compared to control tumors. In addition, MBA-MD-435 tumors exhibited enhanced rate of metastasis to lymph nodes and lung (Skobe et al., 2001). Overexpression of VEGF-C in the MCF-7 human breast carcinoma cell line resulted in increased tumor growth, in absence of angiogenesis, and parallel to lymphangiogenesis, particularly at the tumor periphery (Karpanen et al., 2001; Mattila et al., 2002). Moreover, high rate of metastatic spread to regional lymph nodes was observed different to mice injected with cells transfected with empty vector, and lymphangiogenesis was blocked by a soluble form of VEGFR-3 (VEGFR-3-Ig) (Mattila et al., 2002). Tumors derived from a VEGF-C-overexpressing melanoma cell line gave increased blood and lymphatic vessels and tumor-associated macrophages compared to control tumors (Skobe et al., 2001). Overexpression of VEGF-C in the AZ521 human gastric carcinoma cell line also led to both increased growth of intratumoral lymphatic vessels and metastatic potential of cells when injected into nude mice (Yanai et al., 2001). Overexpression of VEGF-D induced angiogenesis and increased tumor growth compared to controls transfected with the empty vector. VEGF-D also induced extensive lymphangiogenesis and metastatic spread via the lymphatics to draining lymph nodes (Stacker et al., 2001).

10.9.5 Culturing and Purifying of Lymphatic Endothelial Cells

Beginning in 1984, pure cultures of lymphatic endothelium were first isolated by Johnston and Walker from bovine mesenteric lymphatic collecting vessels and by Witte et al. from a patient with a massive cervicomediastinal lymphangioma and another patient with a large retroperitoneal chyle-containing lymphangioma (Way et al., 1987). Lymphatic endothelium was isolated from thoracic duct (Gnepp and Chandler, 1985; Weber et al., 1994a). This vessel is immersed in the periadventitial fat of aorta and can be visualized after injection of Evans blue into its caudal end (Fig. 10.7). Others authors cultured bovine, ovine, rat, and mouse lymphatic endothelium through multiple passages (Yong and Jones, 1991; Leak and Jones, 1993; Borron and Hay, 1994; Weber et al., 1994b). Lymphatic endothelial cells have been grown as monolayer and on microcarrier beads using fetal bovine serum with and without heparin and endothelial cell growth supplement. In each instance, endothelium grew slowly in sheets with a cobblestone morphology, showing characteristic lymphatic-like overlapping cell junctions. Proliferation rate depended on varying growth conditions in primary and passage levels, and there are reports of doubling times as high as a 36–48 hours (Yong and Jones, 1991). Lymphatic explant proliferation and sprouting has also been documented (Nicosia, 1987). Hemangiogenic growth factors, such as VEGF-A and FGF-2, are capable of inducing lymphangiogenic activity in cultures, i.e., lymphatic endothelial cells invade collagen or fibrin gels and form tubules with lumens (Pepper et al., 1994). Indeed, lymphatic endothelial cells may remain pluripotential in vitro and dedifferentiate or transdifferentiate by modifying gene expression common to lymphatic and blood vascular endothelial cells. These findings imply that relatively pure endothelial populations can be isolated from lymphatic vascular tumors and lymphatic collecting vessels and propagated for long periods in vitro while retaining morphologic characteristics similar to blood vascular endothelium. Makinen et al. (2001b) isolated and cultured stable lineages of blood vascular and lymphatic endothelial cells from human dermal microvascular endothelium by using antibodies against the extracellular domain of VEGFR-3. They showed that VEGFR-3 stimulation alone protects the lymphatic endothelial cells from serum deprivation-induced apoptosis and induces their growth and migration.

10.10 MOLECULAR REGULATION OF LYMPHANGIOGENESIS

Major understanding on the development and growth of lymphatic vessels derives from the discovery of a wide array of molecular mediators. To date, the VEGF-C/VEGF-D/VEGFR-3 signaling system is thought to play a central role in the control of these processes in embryogenesis and other conditions.

10.10.1 VEGF-C

VEGF-C is a VEGF isoform closely related to VEGF-D, characterized by the presence of unique amino- and carboxy-terminal extensions flanking the common VEGF-homology domain (Joukov et al., 1996). VEGF-C is synthesized as a precursor protein which undergoes subsequent proteolytic processing reminiscent of the platelet derived growth factor-A and -B (PDGF-A and PDGF-B) chain processing, which suggests an evolutionary relationship (Heldin et al., 1993). The proteolytic processing probably provides a regulatory mechanism which allows a fine tuning of the biological functions of VEGF-C. Unprocessed VEGF-C binds to VEGFR-3, and the stepwise proteolytic processing of VEGF-C generates several VEGF-C forms with increased affinity for VEGFR-3, but only the fully processed VEGF-C is able to activate VEGFR-2 (Joukov et al., 1997). As VEGFR-2 is present in many types of endothelia and VEGF-C is broadly expressed, it may be well that the biosynthesis of VEGF-C as a precursor prevents unnecessary angiogenesis elicited via VEGFR-2 and allows VEGF-C to signal preferentially via VEGFR-3, which is restricted to the venous endothelia during early stages of development and to the lymphatic endothelium during later stages. In certain circumstances, proteolytic processing would release mature VEGF-C, which signals via both VEGFR-3 and VEGFR-2, whose consensual activation may be necessary for full biological response to VEGF-C (Joukov et al., 1997).

VEGF-C may play several functions in the organization of the vascular tree. In VEGF-C deficient mice, endothelial cells commit to the lymphatic lineage but do not sprout to form lymph vessels (Karkkainen et al., 2004). Sprouting was rescued by VEGF-C and VEGF-D but not VEGF-A, indicating VEGFR-3 specificity. The lack of lymphatic vessels resulted in prenatal death due to fluid accumulation in tissues. VEGFR-2-deficient mice die at an earlier stage than

VEGF-A-deficient mice (Dumont et al., 1998). It is thus possible that other VEGFR-2 ligands, including VEGF-C, may compensate for the loss of VEGF-A.

VEGF-C induces lymphangiogenesis in the ears of mice and in the CAM (Enholm et al., 2001; Oh et al., 1997) and lymphatic vessel enlargement in the skin (Jeltsch et al., 1997). VEGF-C has also potent effects on blood vessels because its fully processed form also binds to VEGFR-2 of blood vessels and stimulates angiogenesis (Cao et al., 1998). Furthermore, VEGFR-3 is also expressed on endothelial cells of tumor blood vessels and is thought to play an angiogenic role (Partanen et al., 1999). Compared to VEGF-A, VEGF-C is 4–5 times less potent in the vascular permeability assay (Joukov et al., 1997). VEGF-C mRNA levels are increased by serum and its growth factors PDGF, epidermal growth factor (EGF), and transforming growth factor-β (TGF-β), and by the tumor promoter phorbol myristate 12,13-acetate (PMA) (Enholm et al., 2001). Conversely, hypoxia, Ras oncoprotein, and mutant p53 tumor suppressor do not influence VEGF-C mRNA levels. Interleukin-1β (IL-1β) and tumor necrosis factor-α (TNF-α) stimulate VEGF-C expression in human lung fibroblasts and human umbilical vein endothelial cells (Ristimaki et al., 1998). Furthermore, the antiinflammatory dexamethasone inhibits IL-1β-induced VEGF-C mRNA expression. Hence, VEGF-C seems to be also a mediator of inflammatory reactions (Narko et al., 1999).

However, the biologic effects of VEGF-C are tissue-specific and dependent on the abundance of blood vessels and lymphatics expressing its receptors in a given tissue. Saaristo et al. (2002) characterized the in vivo effects of VEGF-C on blood and lymphatic vessels in the skin and respiratory tract of nude mice: VEGF-C gave a dose-dependent enlargement and tortuosity of veins, which along with the collecting lymphatic vessels were found to express VEGFR-2. Expression of Angl blocked the increased leakiness of the blood vessels induced by VEGF-C, whereas vessel enlargement and lymphangiogenesis were not affected. However, angiogenic sprouting was not observed.

10.10.2 VEGF-D

VEGF-D was first isolated from a differential display screening of murine fibroblast genes from mice carrying a targeted inactivation of the c-fos gene (Orlandini et al., 1996). The identified protein was first named "c-fos-induced growth factor" (FIGF), but later renamed

VEGF-D. After or during secretion, VEGF-D can be proteolytically cleaved at the N- and C-terminal regions of the VHD (Stacker et al., 1999). The processing of VEGF-D is required to produce a growth factor that binds both VEGFR-2 and VEGFR-3 with high affinity. The fully processed VEGF-D binds VEGFR-2 and VEGFR-3 with greater affinity than does unprocessed VEGF-D. The identification of the protease(s) responsible for VEGF-D processing will be important for determining the biological context of the regulation of the receptor affinity and specificity of VEGF-D. VEGF-D is angiogenic in the rabbit cornea assay (Marconcini et al., 1999). In a mouse tumor model, VEGF-D promoted lymphangiogenesis (Achen et al., 1998) and metastatic spread via the lymphatics (Stacker et al., 2001). Lymphatic spread was blocked by a VEGF-D-specific antibody. Achen et al. (2002) analyzed VEGF-D activity in human tumors and a mouse model of metastasis. Tumor vessels positive for VEGF-D were also positive for VEGFR-2 and/or VEGFR-3 but negative for VEGF-D mRNA, indicating that VEGF-D is secreted by tumor cells and subsequently associates with endothelium via receptor-mediated uptake. In the mouse model, VEGF-D synthesized in tumor cells became localized on the endothelium and thereby promoted metastatic spread. Overall data indicate that VEGF-D promotes tumor angiogenesis, lymphangiogenesis, and metastatic spread by a paracrine mechanism.

Byzova et al. (2002) demonstrated that an adenovirus encoding the mature form of human VEGF-D induced predominantly angiogenesis in the rat cremaster muscle, and both angiogenesis and lymphangiogenesis when injected into the epigastric skin. Immunohistochemical analysis of the cremaster muscle demonstrated that adenovirus-induced neovascularization was composed primarily of laminin and VEGFR-2-positive vessels containing red blood cells, thus indicating a predominantly angiogenic response. In the skin model, the adenovirus induced angiogenesis and lymphangiogenesis, as indicated by staining with laminin, VEGFR-2, and VEGFR-3.

These data suggest that the absolute and relative expression levels of VEGFR-2 and VEGFR-3 in endothelium may influence whether VEGF-C/VEGF-D elicit angiogenic or lymphangiogenic effects. A strict demarcation of the effects of VEGFR-2 and VEGFR-3 activation in terms of angiogenesis and lymphangiogenesis is complicated by observations that VEGFR-3 is involved in maintenance of tumor blood vessels in a mouse model (Kubo et al., 2000) and humans (Partanen et al., 1999), and that VEGFR-2 expression is detectable on lymphatic endothelial cells (Kriehuber et al., 2001). Furthermore, the influence of other positive and negative regulators of angiogenesis and lymphangiogenesis may also modulate the biological consequences of VEGF-C and VEGF-D expression.

10.10.3 Vascular Endothelial Growth Factor Receptor-3 (VEGFR-3/Flt4)

VEGFR-3/Flt4 is a highly glycosilated, relatively stable surface tyrosine kinase of approximately 180 kDa. Its cDNA was clone from human erythroleukaemia cells and placental libraries (Aprelikova et al., 1992).

Like VEGFR-1 and VEGFR-2, it is a member of a subfamily of receptor protein tyrosine kinases which are characterized by an extracellular region containing seven immunoglobulin-related domains and an intracellular domain with homology to the platelet-derived growth factor receptor subfamily. Two alternatively spliced isoforms of VEGFR-3 have been described which differ in the length of their cytoplasmic domains and probably in different signaling properties (Borg et al., 1995).

In addition to being expressed on lymphatic endothelial cells, VEGFR-3 has also been detected on cells of the hematopoietic system (Fournier et al., 1995), perhaps reflecting a common stem cell origin for hematopoietic and endothelial cells. Targeted inactivation of the VEGFR-3 gene is lethal in embryo mice (Dumont et al., 1998). The embryo dies because of defective blood vessel and hearth development. This occurs at day 9.5, before the development of the lymphatic system, which emphasizes that VEGFR-3 is expressed in developing blood vessel in embryogenesis and only later becomes sufficiently restricted to the lymphatic system (Kaipainen et al., 1995). Moreover, VEGFR-3 can also be expressed on blood capillaries during tumor angiogenesis (Kubo et al., 2000). Overall, VEGFR-3 is still one of the best and most widely used markers of lymphatic endothelium.

10.11 OTHER REGULATORS OF LYMPHANGIOGENESIS

Other molecules have been implicated in the development and/or maintenance of the lymphatic system and presumably of lymphangiogenesis. Transgenic mice lacking the Ets DNA-binding domain of the transcription factor Net die following birth due to the accumulation of fluid within the thoracic cavity (chylothorax) that is associated with dilated lymphatic vessels (Ayadi et al., 2001). Mice lacking the α_9 integrin subunit also succumb to chylothorax (Huang et al., 2000) and mice deficient in Ang2 have abnormal lymphatic vessels, rescued by Angl (Gale et al., 2002). A protein expressed by lymphatic endothelium, for which a role in lymphatic development is yet to be defined is NRP-2. This is a nontyrosine kinase receptor that binds the collapsin/semaphorin family of axon-guidance molecules and certain members of the VEGF family. It is suggested that NRP-2 is involved in the VEGFR-3-mediated signal transduction at sites where the VEGFR-2 and VEGFR-3 are coexpressed (Karkkainen et al., 2001).

Yuan et al. (2002) showed a selective requirement for NRP-2 in the formation of lymphatic vessels. Loss of NRP-2 function resulted in the absence or severe reduction in the number of small lymphatic vessels and capillaries in all tissues examined. Lymphatic vessel number was decreased as soon as these vessels developed in the embryo and remained reduced or absent until after birth. The lymphatic vessels present in the skin of mutant animals were also located abnormally and sometimes enlarged, suggesting that guidance defects may also occur in the absence of NRP-2 function. Veins, which express lower levels of NRP-2, developed normally in the NRP-2 null animals suggesting that the NRP-2 mutation selectively affected the lymphatic compartment.

10.12 BOTH MECHANISMS MAY CONTRIBUTE TO THE FORMATION OF LYMPHATIC VESSELS

Both of the proposed mechanisms may contribute to the formation of lymphatic system, as suggested by van der Jagt (1932). Wilting et al (2006) provided evidence for a dual origin of the lymphatics of birds from both the veins and scattered mesodermal precursor cells. The main parts of the lymph sacs are derived from adjacent venous vessels, whereas the superficial, dermal lymphatics are derived from local lymphangioblasts. The two compartments fuse to form the potent lymphatic system. Wilting et al (2006) have studied the origin of JLS, the dermal lymphatics and the lymph hearts of the avian embryo. Intravenous injection of DiI-conjugated acetylated low-density-lipoprotein, which labels endothelial cells into day-4 embryos, revealed labeling of the JLS at day 6.5, suggesting a venous origin of the JLS.

Moreover, quail-chick grafting of paraxial mesoderm demonstrated integration of graft-derived lymphatic endothelial cells into superficial parts of the JLS. Finally, the lymphatics of the dermis are directly derived from the dermatomes.

10.13 CONCLUDING REMARKS

Until recently, the lymphatic vessels have received much less attention than blood vessels, despite their importance in medicine. Now, it is well established that the development of the lymphatic system is mainly regulated by the VEGF-C/VEGF-D/VEGFR-3 signaling system. Knowledge about different molecules involved in lymphangiogenesis in health and disease has been accumulated rapidly in the last years.

It has long been known that lymph vessels proliferate during inflammation (Pullinger and Florey, 1937), and inflammatory lymphangiogenesis has now documented in several settings of acute and chronic inflammation (Halin and Detmar, 2008). For example, skin lesions in the chronic inflammatory disease psoriasis show lymphatic hyperplasia (Kuntsfeld et al., 2004), and lymphangiogenesis has also been found to be associated with kidney transplant rejection (Kerjaschki et al., 2004).

Tumor lymphangiogenesis can be stimulated in a variety of experimental cancers by VEGF-C and VEGF-D, and both lymphangiogenesis and the formation of lymph node metastases can be inhibited by antagonists of VEGF-C and VEGF-D (Achen et al., 2005). Numerous studies have shown a direct correlation between VEGF-C and VEGF-D expression by human cancers and tumor metastasis, suggesting that lymphangiogenesis has an important role in promoting metastasis of human tumors (Achen et al., 2005). A cross-talk between angiogenesis and lymphangiogenesis in tumor progression, that is involvement of VEGF-C, VEGF-D, and VEGFR-3 in angiogenesis, and the role played by VEGF-A and angiopoietin-2 (Ang-2) in lymphangiogenesis, respectively, has been established (Scavelli et al., 2004).

Several important questions remain unsolved, regarding the mechanisms by which expression of VEGF-C and VEGF-D is increased in primary tumors and the role of these molecules in promoting lymph node metastasis. Initial preclinical studies have provided promising results in the inhibition of lymphatic tumor metastasis as well as in the stimulation of lymphatic growth in wound healing.

REFERENCES

Achen, M.G., Jeltsch, M., Kukk, E., et al., 1998. Vascular endothelial growth factor D (VEGF-D) is a ligand for the tyrosine kinases VEGF receptor 2 (Flk-1)and VEGF receptor 3 (Flt-4). Proc. Natl. Acad. Sci. U.S.A. 95, 548–553.

Achen, M.G., Williams, R.A., Baldwin, M.E., et al., 2002. The angiogenic and lymphangiogenic factor vascular endothelial growth factor-D exhibits a paracrine mode of action in cancer. Growth Factors 20, 99–107.

Achen, M.G., Mc Coll, B.K., Stacker, S.A., 2005. Focus on lymphangiogenesis in tumor metastasis. Cancer Cell 7, 121–127.

Aprelikova, O., Pajusola, K., Partanen, J., et al., 1992. FLT4, a novel class III receptor tyrosine kinase in chromosome 5q33-qter. Cancer Res. 52, 746–748.

Aselli, G., 1627. De Lactibus sive Lacteis Venis, Quarto Vasorum Mesarai corum Genere novo invento. Medioloni, Milan.

Aukland, K., Reed, R.K., 1993. Interstitial-lymphatic mechanisms in the control of extracellular fluid volume. Physiol. Rev. 73, 1–78.

Ayadi, A., Zheng, H., Sobieszczuk, P., et al., 2001. Net-targeted mutant mice develop a vascular phenotype and up-regulate egr-1. EMBO J. 20, 5139–5152.

Balankura, N.K., 1951. Development of the mammalian lymphatic system. Nature 158, 196–197.

Banerji, S., Ni J, Wang, S.X., et al., 1999. LYVE-1, a new homologue of the CD44 glycoprotein, is a lymph-specific receptor for hyaluronan. J. Cell Biol. 144, 789–801.

Bellman, S., Odén, B., 1959. Regeneration of surgically divided lymph vessels. An experimental study on the rabbit's ear. Acta Chirur. Scan 116, 99–117.

Boardman, K.C., Swartz, M.A., 2003. Interstitial flow as a guide for lymphangiogenesis. Circ. Res. 92, 801–808.

Borg, J.P., deLapeyriere, O., Noguchi, T., et al., 1995. Biochemical characterization of two isoforms of FLT4, a VEGF receptor-related tyrosine kinase. Oncogene 10, 973–984.

Borron, P., Hay, J.B., 1994. Characterization of ovine lymphatic endothelial cells and their interactions with lymphocytes. Lymphology 27, 6–13.

Bowman, C.A., Witte, M.H., Witte, C.L., et al., 1984. Cystic hygroma reconsidered: hamartoma or neoplasm? Primary culture of an endothelial cell line from a massive cervicomediastinal hygroma with bony lymphangiomatosis. Lymphology 17, 15–22.

Breiteneder-Geleff, S., Soleiman, A., Kowalski, H., et al., 1999. Angiosarcomas express mixed endothelial phenotypes of blood and lymphatic capillaries: podoplanin as a specific marker for lymphatic endothelium. Am. J. Pathol. 154, 385–394.

Byzova, T.V., Goldman, C.K., Jankau, J., et al., 2002. Adenovirus encoding vascular endothelial growth factor-D induces tissue-specific vascular patterns in vivo. Blood 99, 4434–4442.

Cao, Y., Linden, P., Farnebo, J., et al., 1998. Vascular endothelial growth factor C induces angiogenesis *in vivo*. Proc. Natl. Acad. Sci. U.S.A. 95, 14389–14394.

Clark, E.R., 1909. Observation on living growing lymphatics in the tail of the frog larva. Anat. Rec. 3, 183–198.

Clark, E.R., 1922. Reaction of experimentally isolated lymphatic capillaries in the tails of amphibian larvae. Anat. Rec. 24, 181–191.

Clark, E.R., Clark, E.L., 1932. Observations on the new growth of lymphatic vessels as seen in transparent chambers introduced into rabbit's ear. Am. J. Anat. 51, 49–87.

Clark, E.R., Clark, E.L., 1933. Further observations on living lymphatic vessels in the transparent chamber in rabbit'ear—their relation to the tissue spaces. Am. J. Anat. 52, 273–305.

Clark, E.R., Clark, E.L., 1937. Observation on living mammalian lymphatic capillaries – their relation to the blood vessels. Am. J. Anat. 60, 253–298.

Clark, E.R., Clark, E.L., 1938. Observation an isolated lymphatic capillaries in the living mammal. Am. J. Anat. 62, 59–92.

de Waal, R.M., van Altena, M.C., Erhard, H., et al., 1997. Lack of lymphangiogenesis in human primary cutaneous melanoma. Consequences for the mechanism of lymphatic dissemination. Am. J. Pathol 150, 1951–1957.

Doyle, D., 2006. William Hewson (1739–74): the father of haematology. Br. J. Haematol. 133, 375–381.

Dumont, D.J., Jussila, L., Taipale, J., et al., 1998. Cardiovascular failure in mouse embryos deficient in VEGF receptor-3. Science 282, 946–949.

Ebata, N., Nodasaka, Y., Sawa, Y., et al., 2001. Desmoplakin as a specific marker of lymphatic vessels. Microvasc. Res. 61, 40–48.

Enholm, B., Karpanen, T., Jeltsch, M., et al., 2001. Adenoviral expression of vascular endothelial growth factor-c induces lymphangiogenesis in the skin. Circ. Res. 88, 623–629.

Fournier, E., Dubreuil, P., Birnbaum, D., et al., 1995. Mutation at tyrosine residue 1337 abrogates ligand-dependent transforming capacity of the FLT4 receptor. Oncogene 11, 921–931.

Gale, N.W., Thurston, G., Hackett, S.F., et al., 2002. Angiopoietin-2 is required for postnatal angiogenesis and lymphatic patterning, and only the latter role is rescued by Angiopoietin-1. Dev. Cell 3, 411–423.

Gerli, R., Ibba, L., Fruschelli, C., 1991. Ultrastructural cytochemistry of anchoring filaments of human lymphatic capillaries and their relation to elastic fibers. Lymphology 24, 105–112.

Gerli, R., Solito, R., Weber, E., et al., 2000. Specific adhesion molecules bind anchoring filaments and endothelial cells in human skin initial lymphatics. Lymphology 33, 148–157.

Gnepp, D.R., Chandler, W., 1985. Tissue culture of human and canine thoracic duct endothelium. In Vitro Cell Dev. Biol. 21, 200–206.

Groger, M., Holnthoner, W., Maurer, D., et al., 2000. Dermal microvascular endothelial cells express the 180-kDa macrophage mannose receptor in situ and in vitro. J. Immunol. 165, 5428–5444.

Halin, C., Detmar, M., 2008. Inflammation, angiogenesis, and lymphangiogenesis. Methods Enzymol. 445, 1–25.

Heldin, C.H., Ostman, A., Westermark, B., 1993. Structure of platelet-derived growth factor: implications for functional properties. Growth Factors 8, 245–252.

Hewson W. Experimental inquires: Part Second. Containing a description of the lymphatic system in the human subject, and in other animals. Together with observations on the lymph, and the changes which it undergoes in some diseases. J. Johnson, No. 72, St Paul's Church Yard, London, 1774.

Hong, Y.K., Harvey, N., Noh, Y.H., et al., 2002. Prox1 is a master control gene in the program specifying lymphatic endothelial cell fate. Dev. Dyn. 225, 351–357.

Huang, X.Z., Wu, J.F., Ferrando, R., et al., 2000. Fatal bilateral chylothorax in mice lacking the integrin alpha9beta1. Mol. Cell Biol. 20, 5208–5215.

Hub, E., Rot, A., 1998. Binding of RANTES, MCP-1, MCP-3, and MIP-1α to cells in human sky. Am. J. Pathol. 152, 749–757.

Huntington, G.S., 1908. The genetic interpretation of the development of mammalian lymphatic system. Anat. Rec. 2, 19–46.

Huntington, G.S., McClure, C.F.W., 1910. The anatomy and development of the jugular lymph sac in the domestic cat (*Felis domestica*). Am. J. Anat. 10, 177–311.

Ikomi, F., Schmid-Schonbein, G.W., 1996. Lymph pump mechanics in the rabbit hind leg. Am. J. Physiol. 271, H173–H183.

Jackson, D.G., Prevo, R., Clasper, S., et al., 2001. LYVE-1, the lymphatic system and tumour lymphangiogenesis. Trends Immunol. 22, 317–321.

Jeltsch, M., Kaipanen, A., Joukov, V., et al., 1997. Hyperplasia of lymphatic vessels in VEGF-C transgenic mice. Science 276, 1423–1425.

Johnston, M.G., Walker, M.A., 1984. Lymphatic endothelial and smooth-muscle cells in tissue culture. In Vitro 20, 566–572.

Joukov, V., Pajusola, K., Kaipainen, A., et al., 1996. A novel vascular endothelial growth factor, VEGF-C, is a ligand for the Flt4 (VEGFR-3) and KDR (VEGFR-2) receptor tyrosine kinases. EMBO J 1996 (15), 290–298.

Joukov, V., Sorsa, T., Kumar, V., et al., 1997. Proteolytic processing regulates receptor specificity and activity of VEGF-C. EMBO J 1997 (16), 3898–3911.

Kaipainen, A., Korhonen, J., Mustonen, T., et al., 1995. Expression of the fms-like tyrosine kinase 4 gene becomes restricted to lymphatic endothelium during development. Proc. Natl. Acad. Sci. U.S.A. 92, 3566–3570.

Kampmeier, O.F., 1912. The value of the injection method in the study of lymphatic development. Anat. Rec. 6, 223–233.

Karkkainen, M., Saaristo, A., Jussila, L., Karila, K., Lawrence, E., Pajusola, K., et al., 2001. A model for gene therapy of human hereditary lymphedema. Proc. Natl. Acad. Sci. U.S.A 98, 12677–12682.

Karkkainen, M., Haiko, P., Sainio, K., Partanen, J., Taipale, J., Petrova, T.V., et al., 2004. Vascular endothelial growth factor C is required for sprouting of the first lymphatic vessels from embryonic veins. Nat. Immunol. 5, 74–80.

Karkkainen, M.J., Haiko, P., Sainio, K., et al., 2004. Lymphatic neoangiogenesis in human kidney transplants is associated with immunologically active lymphocytic infiltrates. J. Am. Soc. Nephrol. 15, 603–612.

Karpanen, T., Egeblad, M., Karkkainen, M.J., et al., 2001. Vascular endothelial growth factor C promotes tumour lymphangiogenesis and intralymphatic tumour growth. Cancer Res. 2001 (61), 1786–1790.

Kato, S., 1900. Enzyme-histochemical identification of lymphatic vessels by light and backscattered image scanning electron microscopy. Stain Technol. 65, 131–137.

Kerjaschki, D., Regele, H.M., Moosberger, I., et al., 2004. Lymphatic neoangiogenesis in human kidney transplants is associated with immunologically active lymphocytic infiltrates. J. Am. Soc. Nephrol. 15, 603–612.

Kowalczyk, A.P., Navarro, P., Dejana, E., et al., 1998. VE cadherin and desmoplakin are assembled into dermal microvascular endothelial intercellular junctions: a pivotal role for plakoglobin in therecruitment of desmoplakin to intercellular junctions. J. Cell Sci. 1998 (11), 3045–3057.

Kriehuber, E., Breiteneder-Geleff, S., Groeger, M., et al., 2001. Isolation and characterization of dermal lymphatic and blood endothelial cells reveal stable and functionally specialized cell lineages. J. Exp. Med. 2001 (194), 797–808.

Kubo, H., Otsuki, Y., Magari, S., et al., 1990. Comparative study of lymphatics and blood vessels. Acta Histochem. Cytochem. 23, 621–636.

Kubo, H., Fujiwara, T., Jussila, L., et al., 2000. Involvement of vascular endothelial growth factor receptor-3 in maintenance of integrity of endothelial cell lining during tumour angiogenesis. Blood 2000 (96), 546–553.

Kunstfeld, R., Hirakawa, S., Hong, Y.K., et al., 2004. Induction of cutaneous delayed-type hypersensitivity reactions in VEGF-A transgenic mice results in chronic skin inflammation associated with persistent lymphatic hyperplasia. Blood 104, 1048–1057.

Le Douarin, N.M., 1969. Particularités du noyau interphasique chez la caille japonaise (*Coturnix coturnix japonica*). Utilisation de ces particularitès comme "marquage biologique" dans les recherches sur les interactions tissulaires et les migrationes cellulaires au cours de l'ontogenése. Bull. Biol. Fr. Belg. 103, 435–452.

Leak, L.V., 1970. Electron microscopic observations on lymphatic capillaries and the structural components of the connective tissue-lymph interface. Microvasc. Res. 2, 361–391.

Leak, L.V., Jones, M., 1993. Lymphatic endothelium isolation characterization and long term culture. Anat. Rec. 236, 641–652.

Leak, L.V., Jones, M., 1994. Lymphangiogenesis in vitro: formation of lymphatic capillary-like channels from confluent monolayers of lymphatic endothelial cells. In Vitro Cell Dev. Biol. Anim. 30, 512–518.

Linehan, S.A., Martinez-Pomares, L., da Silva, R.P., et al., 2001. Endogenous ligands of carbohydrate recognition domains of the mannose receptor in murine macrophages, endothelial cells and secretory cells; potential relevance to inflammation and immunity. Eur. J. Immunol. 6, 1857–1866.

Leu, A.J., Berk, D.A., Yuan, F., et al., 1994. Flow velocity in the superficial lymphatic network of the mouse tail. Am. J. Physiol. 267, H1507–H1513.

Lewis, F.T., 1905. The development of the lymphatic system in rabbits. Am. J. Anat. 5, 95–111.

Lewis, F.T., 1909. On the cervical veins and lymphatics in four human embryos. With an interpretation of anomalies of the subclavian and jugular veins in the adult. Am. J. Anat. 9, 33–43.

Magari, S., Asano, S., 1978. Regeneration of the deep cervical lymphatics-light and electron microscopic observations. Lymphology 11, 57–61.

Magari, S., 1987. Comparison of fine structure of lymphatics and blood vessels in normal conditions and during embryonic development and regeneration. Lymphology 20, 189–195.

Makinen, T., Jussila, L., Veikkola, T., et al., 2001a. Inhibition of lymphangiogenesis with resulting lymphedema in transgenic mice expressing soluble VEGF receptor-3. Nat. Med. 2, 199–205.

Makinen, T., Veikkola, T., Mustjoki, S., et al., 2001b. Isolated lymphatic endothelial cells transduce growth, survival and migratory signals via the VEGF-C/D receptor VEGFR-3. EMBO J 20, 4762–4773.

Mandriota, S.J., Jussila, L., Jeltsch, M., et al., 2001. Vascular endothelial growth factor-C mediated lymphangiogenesis promotes tumour metastasis. EMBO J 20, 672–682.

Marconcini, L., Marchio, S., Morbidelli, L., et al., 1999. c-fos-induced growth factor/vascular endothelial growth factor D induces angiogenesis in vivo and in vitro. Proc. Natl. Acad. Sci. U.S.A. 1999 (96), 9671–9676.

Matsui, K., Breitender-Geleff, S., Soleiman, A., et al., 1999. Podoplanin, a novel 43-kDa membrane protein, controls the shape of podocytes. Nephrol. Dial. Transplant. Suppl. 14, 9–11.

Mattila, M.M., Ruohola, J.K., Karpanen, T., et al., 2002. VEGFC induced lymphangiogenesis is associated with lymph node metastasis in orthotopic MCF-7 tumours. Int. J. Cancer 98, 946–951.

McClure, C., 1908. The development of the thoracic and right lymphatic ducts in the domestic cat (*Felis domestica*). Anat. Anz. 32, 533–543.

Miller, A.M., 1912. The development of the jugular lymph sac in birds. Am. J. Anat. 12, 473–490.

Mouta Carreira, C., Nasser, S.M., di Tomaso, E., et al., 2001. LYVE-1 is not restricted to the lymph vessels: expression in normal liver blood sinusoids and down-regulation in human liver cancer and cirrhosis. Cancer Res. 61, 8079–8084.

Narko, K., Enholm, B., Makinen, T., et al., 1999. Effect of inflammatory cytokines on the expression of the vascular endothelial growth factor-C. Int. J. Exp. Pathol. 80, 109–112.

Negrini, D., Passi, A., de Luca, G., et al., 1996. Pulmonary interstitial pressure and proteoglycans during development of pulmonary edema. Am. J. Physiol. 270, 2000–2007.

Nibbs, R.J.B., Wylie, S.M., Pragnell, I.B., et al., 1997. Cloning and characterization of a novel murine β-chemokine receptor, D6: comparison to three other related macrophage inflammatory protein-1α receptors, CCR-1, CCR-3, and CCR-5. J. Biol. Chem. 272, 12495–12504.

Nibbs, R.J., Kriehuber, E., Ponath, P.D., et al., 2001. The β-chemokine receptor D6 is expressed by lymphatic endothelium and a subset of vascular tumours. Am. J. Pathol. 158, 867–877.

Nicosia, R.F., 1987. Angiogenesis and the formation of lymphatic like channels in cultures of thoracic duct. In Vitro Cell Dev. Biol. 23, 167–174.

Odén, B., 1960. A micro-lymphangiographic study of experimental wounds healing by second intention. Acta Chir. Scand. 120, 100–114.

Oh, S.J., Jeltsch, M.M., Birkenhager, R., et al., 1997. VEGF and VEGF-C: specific induction of angiogenesis and lymphangiogenesis in the differentiated avian chorioallantoic membrane. Dev. Biol. 188, 96–109.

Orlandini, M., Marconcini, L., Ferruzzi, R., et al., 1996. Identification of a c-fos-induced gene that is related to the platelet-derived growth factor/vascular endothelial growth factor family. Proc. Natl. Acad. Sci. U.S.A. 93, 11675–11680.

Paavonen, K., Puolakkainen, P., Jussila, L., et al., 2000. Vascular endothelial growth factor receptor-3 in lymphangiogenesis in wound healing. Am. J. Pathol. 156, 1499–1504.

Papoutsi, M., Tomarev, S.I., Eichmann, A., et al., 2001. Endogenous origin of the lymphatics in the avian chorioallantoic membrane. Dev. Dyn. 222, 238–251.

Pardanaud, L., Altmann, C., Kitos, P., et al., 1987. Vasculogenesis in the early quail blastodisc as studied with a monoclonal antibody recognizing endothelial cells. Development 100, 339–349.

Partanen, T.A., Alitalo, K., Miettinen, M., 1999. Lack of lymphatic vascular specificity of vascular endothelial growth factor receptor-3 in 185 vascular tumours. Cancer 86, 2406–2412.

Pepper, M.S., Wasi, S., Ferrara, N., et al., 1994. In vitro angiogenic and proteolytic properties of bovine lymphatic endothelial cells. Exp. Cell Res. 210, 298–305.

Prevo, R., Banerji, S., Ferguson, D.J., et al., 2001. Mouse LYVE-1 is an endocytic receptor for hyaluronan in lymphatic endothelium. J. Biol. Chem 276, 19420–19430.

Pullinger, D.B., Florey, H.W., 1937. Proliferation of lymphatics in inflammation. J. Pathol. Bacteriol. 45, 157–170.

Ranvier, L., 1895. Développement des vaisseaux lymphatiques. C. R. Acad. Sci. 121, 1105–1109.

Reading, P.C., Miller, J.L., Anders, E.M., 2000. Involvement of the mannose receptor in infection of macrophages by influenza virus. J. Virol. 74, 5190–5197.

Reichert, F.L., 1926. The regeneration of the lymphatics. Arch. Surg. 13, 871–881.

Ribatti, D., Nico, B., Vacca, A., et al., 2001. Chorioallantoic membrane capillary bed: a useful target for studying angiogenesis and anti-angiogenesis in vivo. Anat. Rec. 264, 317–324.

Ribatti, D., 2009. William Harvey and the discovery of the circulation of the blood. J. Angiogenesis. Res. 1, 3.

Ristimaki, A., Narko, K., Enholm, B., et al., 1998. Proinflammatory cytokines regulate expression of the lymphatic endothelial mitogen vascular endothelial growth factor-C. J. Biol. Chem. 273, 8413–8418.

Ryan, T.J., Mortimer, P.S., Jones, R.L., 1986. Lymphatics of the skin. Int. J. Dermatol. 25, 411–419.

Saaristo, A., Veikkola, T., Enholm, B., et al., 2002. Adenoviral VEGF-C overexpression induces blood vessel enlargement, tortuosity, and leakiness but no sprouting angiogenesis in the skin or mucous membranes. FASEB J 16, 1041–1049.

Sabin, F.R., 1902. On the origin of the lymphatic system from the veins and the development of the lymph hearts and thoracic duct in the pig. Am. J. Anat. 1, 367–391.

Sabin, F.R., 1904. On the development of the superficial lymphatics in the skin of the pig. Am. J. Anat. 3, 183–195.

Sabin, F.R., 1909. The lymphatic system in human embryos, with consideration of the system as a whole. Am. J. Anat. 9, 43–91.

Sabin, F.R., 1911. A critical study of the evidence presented in several recent articles on the development of the lymphatic system. Anat. Rec. 1911 (5), 417–446.

Sacchi, G., Weber, E., Aglianò, M., et al., 1997. The structure of superficial lymphatics in the human thigh: precollectors. Anat. Rec. 247, 53–62.

Sandison, J.C., 1924. A new method for the microscopic study of living growing tissues by introduction of a transparent chamber in the rabbit's ear. Anat. Rec. 28, 281–287.

Sawa, Y., Shibata, K., Braithwaite, M.W., et al., 1999. Expression of immunoglobulin superfamily members on the lymphatic endothelium of inflamed human small intestine. Microvasc. Res. 57, 100–106.

Scavelli, C., Vacca, A., Di Pietro, G., et al., 2004. Crosstalk between angiogenesis and lymphangiogenesis in tumor progression. Leukemia 18, 1054–1058.

Schacht, V., Ramirez, M.I., Hong, Y.K., et al., 2003. T1 alpha/podoplanin deficiency disrupts normal lymphatic vasculature formation and causes lymphedema. EMBO J. 22, 3546–3556.

Schlingemann, R.O., Dingjan, G.M., Emeis, J.J., et al., 1985. Monoclonal antibody PAL-E specific for endothelium. Lab. Invest. 52, 71–76.

Schmelz, M., Moll, R., Kuhn, C., et al., 1994. Complexus adhaerentes, a new group of desmoplakin containing junctions in endothelial cells: II Different types of lymphatic vessels. Differentiation 57, 97–117.

Schneider, M., Othman-Hassan, K., Christ, B., et al.,Wilting, J. 1999. Lymphangioblasts in the avian wing bud. Dev. Dyn. 1999 (216), 311–319.

Skobe, M., Hamberg, L.M., Hawighorst, T., et al., 2001. Concurrent induction of lymphangiogenesis, angiogenesis, and macrophage recruitment by vascular endothelial growth factor-C in melanoma. Am. J. Pathol. 159, 893–903.

Slavin, S.A., Van den Abbeele, A.D., Losken, A., et al., 1999. Return of lymphatic function after flap transfer for acute lymphedema. Ann. Surg. 229, 421–427.

Sleeman, J.P., Krishnan, J., Kirkin, V., et al., 2001. Markers for the lymphatic endothelium: in search of the holy grail? Microsc. Res. Tech. 55, 61–69.

Solito, R., Alessandrini, C., Fruschelli, M., et al., 1997. An immunological correlation between the anchoring filaments of initial lymph vessels and the neighboring elastic fibers: a unified morphofunctional concept. Lymphology 30, 194–202.

Stacker, S.A., Stenvers, K., Caesar, C., et al., 1999. Biosynthesis of vascular endothelial growth factor-D involves proteolytic processing which generates non-covalent homodimers. J. Biol. Chem. 274, 32127–32136.

Stacker, S., Caesar, C., Baldwin, M., et al., 2001. VEGF-D promotes the metastatic spread of tumour cells via the lymphatics. Nat. Med. 7, 186–191.

van der Putte, S.C., 1975. The development of the lymphatic system in man. Adv. Anat. Embryol. Cell Biol. 51, 3–60.

Stanton, A.W., Kadoo, P., Mortimer, P.S., et al., 1997. Quantification of the initial lymphatic network in human normal forearm skin using fluorescence microlymphography and stereological methods. Microvasc. Res. 54, 156–163.

Tan, Y., 1998. Basic fibroblast growth factor-mediated lymphangiogenesis of lymphatic endothelial cells isolated from dog thoracic ducts: effects of heparin. Jpn. J. Physiol. 48, 133–141.

Toole, B.P., 1990. Hyaluronan and its binding proteins, the hyaladherins. Curr. Opin. Cell Biol. 2, 839–844.

Van der Jagt, E.R., 1932. The origin and development of the anterior lymph sacs in the sea-turthe (*Thalassochelys caretta*). Quart. J. Microsc. Sci. 75, 151–165.

Veikkola, T., Jussila, L., Makinen, T., et al., 2001. Signalling via vascular endothelial growth factor receptor-3 is sufficient for lymphangiogenesis in transgenic mice. EMBO J. 20, 1223–1231.

Way, D., Hendrix, M., Witte, M., et al., 1987. Lymphatic endothelial cell line (CH3) from a recurrent retroperitoneal lymphangioma. In Vitro 23, 647–652.

Weber, E., Lorenzoni, P., Lozzi, G., et al., 1994a. Culture of bovine thoracic duct endothelial cells. In Vitro Cell Dev. Biol. Anim. 30, 287–288.

Weber, E., Lorenzoni, P., Lozzi, G., et al., 1994b. Differentiation between blood and lymphatic endothelium: bovine blood and lymphatic large vessels and endothelial cells in culture. J. Histochem. Cytochem 42, 1109–1115.

Weber, E., Rossi, A., Solito, R., et al., 2002. Focal adhesion molecules expression and fibrillin deposition by lymphatic and blood vessel endothelial cells in culture. Microvasc. Res. 64, 47–55.

Wigle, J.T., Harvey, N., Detmar, M., et al., 2002. An essential role for Prox1 in the induction of the lymphatic endothelial cell phenotype. EMBO J. 21, 1505–1513.

Wilting, J., Papoutsi, M., Othman-Hassan, K., et al., 2001. Development of the avian lymphatic system. Microsc. Res. Tech. 55, 81–91.

Wilting, J., Aref, Y., Huang, R., et al., 2006. Dual origin of avian lymphatics. Dev. Biol. 292, 165–173.

Witmer, A.N., van Blijswijk, B.C., Dai, J., et al., 1991. A comparison study of cultured vascular and lymphatic endothelium. Exp. Pathol. 42, 11–25.

Yanai, Y., Furuhata, T., Kimura, Y., et al., 2001. Vascular endothelial growth factor C promotes human gastric carcinoma lymph node metastasis in mice. J. Exp. Clin. Cancer Res. 20, 419–428.

Yong, L.C., Jones, B., 1991. A comparison study of cultured vascular and lymphatic endothelium. Exp. Pathol. 42, 11–25.

Yuan, L., Moyon, D., Pardanaud, L., et al., 2002. Abnormal lymphatic vessel development in neuropilin 2 mutant mice. Development 129, 4797–4806.

FURTHER READING

Abtahian, F., Guerriero, A., Sebzda, E., et al., 2003. Regulation of blood and lymphatic vascular separation by signaling protein SLP-76 and Syk. Science 299, 247–251.

Buttler, K., Kreysing, A., von Kaisenberg, C.S., et al., 2006. Mesenchymal cells with leukocyte and lymphendothelial characteristics in murine embryos. Dev. Dyn. 235, 1554–1562.

He, L., Papoutsi, M., Huang, R., et al., 2003. Three different fates of cells migrating from somites into the limb bud. Anat. Embryol. (Berl.) 207, 29–34.

Jackson, D.G., Talikka, M., Rauvala, H., et al., 2004. Vascular endothelial growth factor C is required for sprouting of the first lymphatic vessels from embryonic veins. Nat. Immunol. 5, 74–80.

Ny, A., Koch, M., Schneider, M., et al., 2005. A genetic *Xenopus laevis* tadpole model to study lymphangiogenesis. Nat. Med. 11, 998–1004.

Petrova, T.V., Makinen, T., Makela, T.P., et al., 2002. Lymphatic endothelial reprogramming of vascular endothelial cells by the Prox-1 homeobox transcription factor. EMBO J. 21, 4593–4599.

Wigle, J.T., Oliver, G., 1999. Proxl function is required for the development of the murine lymphatic system. Cell 98, 769–778.

Wilting, J., Schneider, M., Papoutsi, M., et al., 2000. An avian model for studies of embryonic lymphangiogenesis. Lymphology 3, 81–94.

Printed in the United States
By Bookmasters